Applied Multivariate Data Analysis

APPLIED MULTIVARIATE DATA ANALYSIS

Brian S Everitt
Professor of Statistics in Behavioural Science, Institute of Psychiatry, London

and

Graham Dunn
Head, Department of Biostatistics and Computing, Institute of Psychiatry, London

Edward Arnold
A division of Hodder & Stoughton
LONDON   MELBOURNE   AUCKLAND

© 1991 B.S. Everitt and G. Dunn

First published in Great Britain 1991

British Library Cataloguing in Publication Data

Everitt, B.S.
  Applied multivariate data analysis.
  I. Title II. Dunn, G.
  519.5076

  ISBN 0-340-54529-1

Printed and bound in Great Britain for Edward Arnold, a division
of Hodder and Stoughton Limited, Mill Road, Dunton Green,
Sevenoaks, Kent TN13 2YA, by the University Press, Cambridge.

Foreword

The present text arises from an extensive revision of our previous book *Advanced Methods of Data Exploration and Modelling*. Since so much new material is included, particularly in those sections dealing with linear models and latent variable models, we thought it appropriate to regard the work as new rather than simply a second edition. Consequently we have taken the opportunity to give the book a more appropriate title.

Dedication

To my sisters.

Preface

The developments of multivariate analysis from the 1930s until recent years have concentrated on statistical methods and associated distribution theory tied to well-defined models, usually based on the multivariate normal distribution. Consequently many existing textbooks of statistics and data analysis, written at the same level as *Applied Multivariate Data Analysis*, stress hypothesis testing at the expense of what is now generally referred to as *data exploration* which, according to Andrews (1978), is 'the manipulation, summarisation and display of data to make them more comprehensible to human minds, thus uncovering underlying structure in the data and detecting important departures from that structure.'

It is the exploratory phase of data analysis leading to the generation of hypotheses and speculation, that we stress in this text, with significance tests most often being used as diagnostic aids rather than as formal decision making procedures. We hope that in this way we will convey to readers a feel for the analysis of real data and concentrate their thinking on such important questions as 'what do the data show us in the midst of their apparent chaos? and 'how can the data be summarised and represented most informatively?', since it is clear that in many areas, particularly the social and behavioural sciences, that scientific activity is dominated by such questions. Very rarely can one hope to discover natural laws or actually 'explain' anything. The themes of discovery of patterns, creation of models and the assessment of their adequacy are therefore those that have guided the writing of this book.

The text itself comprises four largely self-contained parts. Part I considers particular approaches to data anaysis and the mathematical background necessary for understanding the remainder of the text, and should be studied by all readers. Part II covers the exploration of multivariate data, while Parts III and IV introduce linear models for manifest (observable) and latent variables respectively. Parts III and IV can be read without any knowledge of Part II although it might be useful for the reader to study Chapters 4, 7 and 8 before attempting Part IV. The exercises given at the end of each chapter are primarily practical, although a small number of theoretical problems are included and the solutions to some of these appear in Appendix B. We assume throughout that readers have a firm grasp of basic statistics and mathematics.

We would like to thank those authors and publishers who have given permission for their material to be reproduced or adapted for use in this book, and those colleagues at the Institute of Psychiatry who have provided data for us to include as examples. Finally, we must express our gratitude to Mr. James Griffin of Edward Arnold for support during the writing of this book, and to Ms. Marie Dyer for typing the manuscript.

B.S. Everitt
G. Dunn
Institute of Psychiatry, London, 1990

Contents

PART I
APPROACHES TO ANALYSING DATA

The first two chapters cover essentially introductory material. Chapter 1 reviews the authors' approach to the analysis of complex social and behavioural data, and stresses the importance of model building in both the exploratory and hypothesis testing stages of an investigation. Chapter 2 describes briefly a few mathematical and statistical topics particularly relevant to the remainder of the text.

1

Data and Statistics

1.1 Introduction

The methods used in the systematic pursuit of knowledge are the same, or at least very similar, in all branches of science. They involve the recognition and formulation of problems, the collection of relevant empirical data through passive observation or through experiment, and often the use of mathematical and statistical analysis to explore relationships in the data or to test specific hypotheses about the observations. But there are special problems and difficulties in some disciplines which do not perhaps exist in others. For example, the complexity and ambiguity of some aspects of human behaviour create major problems for the behavioural scientist when trying to draw reliable and valid inferences which may not exist for his counterpart working in physics or chemistry. Consequently it has long been recognised by such scientists that they will generally need to employ relatively more sophisticated and complex analytical tools to investigate their data. In some cases these tools have been supplied by statisticians, in others they have been developed first by the subject matter researchers themselves. A classic example of the latter is *factor analysis*, the basic concepts of which originated with Spearman as early as 1904 and were later extended by Burt and Thompson in the 1920s and 1930s. It was only some time after this that a statistician, Lawley (1941) considered the problem in a more formal way and applied maximum likelihood methods to the estimation of the parameters in the factor analysis model, and only during the last two decades that suitable algorithms have been developed for finding such estimates.

Routine use of many methods of statistical analysis has had to await the dramatic revolution in data analysis brought about by the development and increasing availability of the electronic computer. Thirty years ago, a psychologist wishing to apply factor analysis to a correlation matrix would have been faced with the daunting task of several days or even weeks of hand calculation. Today the equivalent analysis can be performed in a matter of minutes using a relatively inexpensive personal computer. Well-documented statistical packages such as SPSS, SAS, BMDP, GLIM and CLUSTAN enable even the most complicated of procedures to be performed with ease *often by the mathematically and statistically unsophisticated.* This is, of course, not without its problems, and probably results in a number of worthless research findings in some areas. Computers and associated statistical packages are however here to stay and it would be unrealistic to ignore the great contribution they have made in removing the necessity for the investigator to be involved in large amounts of tedious arithmetic. But computers offer far more than merely a means of short-cutting laborious calculations; their ability to produce graphical and geometrical representations of data, and to operate in an interactive mode, make them an essential tool at almost all stages in the investigation of a set of data. (On the other hand it should not be forgotten that in some circumstances there may be simple 'pencil and paper' techniques that produce a more rewarding result than pages and pages of computer printout.)

In this text we will assume that the majority of analyses undertaken by today's research workers are performed with the assistance of a computer and one or other statistical package, and we will take this as a justification for providing only very few details of the arithmetical calculations involved in the techniques to be described. Our main concern will be to concentrate on highlighting when particular methods are applicable and on the correct interpretation of results. In this way we hope to raise the likelihood of researchers drawing valid conclusions from their analyses.

1.2   Types of Data

The data with which we shall be primarily concerned in later chapters consists of a series of measurements or observations made on a number of *subjects, patients, objects* or other *entities* of interest. Such data may be represented in a general way by a matrix **X** given by

$$\mathbf{X} = \begin{pmatrix} x_{11} & x_{12} & \cdots & x_{1p} \\ x_{21} & x_{22} & \cdots & x_{2p} \\ \vdots & \vdots & \ddots & \vdots \\ x_{n1} & x_{n2} & \cdots & x_{np} \end{pmatrix} \tag{1.1}$$

whose typical element x_{ij} represents the value of the jth variable for the ith individual. The number of individuals under investigation is represented by n and the number of measurements taken on each individual by p. Table 1.1 gives a hypothetical example of such a *multivariate data matrix*.

In many cases the variables measured will be of different types each corresponding to one of the following:

(1)   *Nominal* - unordered categorical variables.
(2)   *Ordinal* - where there is ordering but no implication of distance between the different points on the scale.
(3)   *Interval* - where there are equal differences between successive points in the scale but where the zero point is abitrary.
(4)   *Ratio* - the highest level of measurement, where one can compare differences in scores as well as the relative magnitude of scores.

In Table 1.1, for example, sex and depression are nominal scale variables, self-perception of health is ordinal, so probably is IQ, while age has the

Table 1.1 Data Matrix for a hypothetical sample of 10 individuals

Indiv.	Sex	Age	IQ	Depression	SP Health	Weight (lb)
1	Male	21	120	Yes	Very Good	150
2	Male	43	N.K.	No	Very Good	160
3	Male	22	135	No	Average	135
4	Male	86	150	No	Very Poor	140
5	Male	60	92	Yes	Good	110
6	Female	16	130	Yes	Good	110
7	Female	N.K.	150	Yes	Very Good	120
8	Female	43	N.K.	Yes	Average	120
9	Female	22	84	No	Average	105
10	Female	80	70	No	Good	100

N.K. = Not Known

properties of a ratio scale. As we shall see later, different methods of analysis are generally suitable for different types of variables. In some cases the variables recorded may fall into two groups, namely *response* and *explanatory* variables. This situation arises when we are interested in investigating the effect of one set of variables (the explanatory variables) on another set (the response variables). For example, in Table 1.1 we might be interested in assessing the effect of age, weight and sex on individuals' perceptions of their own health.

The data in Table 1.1 also illustrate a problem often met with in practice — that of *missing values*. Age has not been recorded for individual number 7, and no value of IQ is given for individual 2.

Missing observations arise for a variety of reasons, and it is important to put some effort into discovering why an observation is missing. If an investigator has simply forgotten to record all the necessary variables it may be possible to eliminate the missing values by simply contacting the individuals involved. In cases where a respondent in a survey refuses to answer all the questions this will, of course, not be possible. Missing values can cause problems for many of the methods of analysis to be described in this text. If only one or two such values occur the missing observations can often be replaced by estimated values (for example, appropriate means), without causing too much concern, although more complex approaches to estimating missing values have been developed, many of which are described in Little and Rubin (1987). If a substantial number of entries in a multivariate data matrix are missing the whole question of statistical analysis becomes problematical.

1.3 Exploratory and Confirmatory Techniques

It is often suggested that it is helpful to recognise that the analysis of data involves two separate stages. The first, particularly in new areas of research, involves data *exploration* in an attempt to recognise any non-random pattern or structure requiring explanation. At this stage, finding the question is often of more interest than seeking the subsequent answer, the aim of this part of the analysis being to generate possibly interesting hypotheses for later study. Here, formal models designed to yield specific answers to rigidly defined questions are not required. Instead methods are sought which allow possibly unanticipated patterns in the data to be detected, opening up a wide range of alternative explanations. Such techniques are generally characterised by their emphasis on the importance of visual displays and graphical representations and by the lack of any associated stochastic model, so that the questions of the statistical significance of results are hardly ever of importance.

A confirmatory analysis becomes possible once a research worker has some well-defined hypothesis in mind. It is here that some type of statistical significance test might be considered. Such tests are well known to behavioural scientists and, although their misuse has often brought them into some disrepute, they remain of considerable importance.

In this text we have loosely divided the methods of analysis we shall discuss into exploratory (Chapters 3–6) and confirmatory (Chapters 7–14), but this division should not be regarded as much more than a convenient arrangement of the material to be presented, since any sensible investigator will realise the need for both exploratory and confirmatory techniques, and many methods will often be useful in both roles. Linear regression (see Chapter 8), for example, can be both an aid in the search for structure in a set of data via graphical analysis and examination of residuals, and as a formal model for the construction of significance tests. Factor analysis (see Chapter 13) and covariance structure models (Chapter 14) can also be used as both exploratory and confirmatory techniques.

Perhaps recent attempts to rigidly divide data analysis into exploratory and confirmatory parts have been misplaced, and what is really important is that research workers should have a flexible and pragmatic approach to the analysis of their data, with sufficient expertise to enable them to choose the appropriate analytical tool and use it correctly.

1.4   The Role of Models

The aim of many of the techniques described in this text is a simplified description of the structure of the observations by means of what is usually referred to as a *model*. This model can range from a fairly imprecise verbal description to a geometrical representation or mathematical equation. The latter is a precise description of what the investigator visualises as occurring in the population of interest, and in many cases may be the basis of a formal test of significance. Geometrical or graphical representations of the data may often, however, have a more powerful impact. The purpose of building a model is to provide the simplest description of the population being studied *that is consistent with the data*. Of course, if one makes the models complex enough they are bound to fit, but a complicated model may have less explanatory power than a simple, more elegant one. Also, the simpler the model, the easier it is to interpret.

In this text then a model is considered to be a way of simplifying and portraying the structure in a set of data, and does not necessarily imply any causal links or explicitly invoke any causal mechanisms. In the physical sciences, of course, the models used often do imply a mechanism, and are often suggested by a particular theoretical framework. A mathematical description or model of planetary movements, for example, can be derived from Newton's laws of motion. The social and behavioural scientist, however, is very rarely

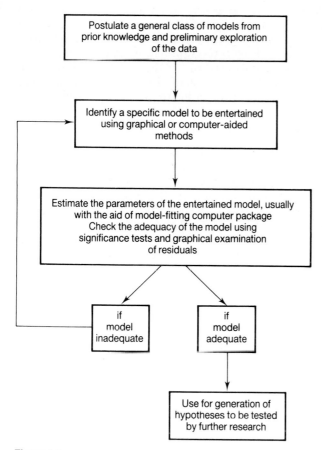

Figure 1.1

in the situation where causal models can be generated from *a priori* theories. Indeed, even when the data can be modelled in this way, the result will very likely have only a restricted area of applicability. Consequently our concern will be to show workers in such disciplines how to simplify and describe their data, and how to construct simple *empirical* models rather than mechanistic ones.

The model-building process is summarised in Figure 1.1. At the first stage a knowledge of the way the data have been collected, the scales of measurement used, and the results of some preliminary exploratory analyses allow the investigator to postulate a very general class of models that might prove useful.

Exploratory techniques remain important in the second stage, in identifying a particular model that can be entertained by the investigator. Then one moves on to the so-called confirmatory stage in which the parameters of the model are estimated and its goodness-of-fit tested. In addition a graphical or geometrical representation of the supposed structure in the data may be produced. Lastly a series of diagnostic procedures may be used to assess the adequacy of the model, or to check whether the underlying assumptions of the model are sensible and justified, by a re-examination of the original data. If it is decided that the model is not justified, or does not fit the data well enough, then a return to an earlier phase is called for, with a renewed search for alternative models. Such a procedure can be summarised in terms of the equation

$$\text{data} = \text{model} + \text{residual}$$

The model is the underlying, simplified structure of set of data that is considered at any particular stage. The residual represents the difference between the model and the observed data points. What is desirable is that the residual should contain no additional pattern or structure. If the residual does contain additional structure then the model associated with the data needs refinement and the process should continue until all parts of this strucure can be assumed to be contained within the model. When constructing a model, however, the following quotation from Box(1965) should be kept firmly in mind.

All models are wrong, but some are useful.

Most of this text is written from the point of view that there are no rules or laws of scientific inference (Feyerabend, 1975). This implies that we see both exploratory and confirmatory techniques as methods of searching for structure in data rather than as decision-making procedures. For this reason we do not stress the values of significance levels, but merely use them as criteria to guide the modelling process. We believe that in scientific research it is the skilful interpretation of evidence and subsequent development of 'hunches' that are important rather than a rigid adherence to a formal set of decision rules associated with significance tests.

2

Mathematical and Statistical Background

2.1 Introduction

As we mentioned in the preface, we are assuming that readers of the text have a firm grasp of basic statistics and mathematics, including some familiarity with matrix algebra. Although this should ensure that most of the material in the remaining chapters can be readily understood, we will in this chapter briefly cover the mathematical and statistical concepts most relevant to the remainder of the text.

2.2 Matrices

Examples of matrices have already been given in Chapter 1. They are simply square or rectangular arrangements of numbers and symbols, and we will assume that readers are familiar with the basic concepts of their algebra such as addition, multiplication and inversion. (Readers not familiar with these topics are referred to Maxwell, 1977, Chapter 3.) Here we shall concentrate on describing a number of other aspects of matrices that are particularly relevant at various points in this text.

2.2.1 Rank of a Matrix

Consider the three equations

$$2x + y - 5 = 0$$
$$x - 2y = 0 \qquad (2.1)$$
$$3x - y - 5 = 0$$

All three equations are satisfied by the values $x = 2$ and $y = 1$, so that they are *consistent*. The equations can be written as follows:

$$\mathbf{Az} = \mathbf{0}$$

where

$$\mathbf{A} = \begin{pmatrix} 2 & 1 & -5 \\ 1 & -2 & 0 \\ 3 & -1 & -5 \end{pmatrix}$$

$$\mathbf{z} = \begin{pmatrix} x \\ y \\ 1 \end{pmatrix} \tag{2.2}$$

$$\mathbf{0} = \begin{pmatrix} 0 \\ 0 \\ 0 \end{pmatrix}$$

Now, since only two of the original three equations are necessary for solving for x and y, the matrix $\mathbf{A}$ contains redundant information; its rows (or columns) are not all independent.

The number of independent rows (or columns) of a matrix is called the *rank* of the matrix. The rank of a matrix can never exceed the smaller of its two dimensions, i.e. if $r(\mathbf{A})$ is the rank of matrix $\mathbf{A}$ of order $m \times n$ (which we shall sometimes write as $\mathbf{A}(m \times n)$, then

$$r(\mathbf{A}) \leq \min(m, n) \tag{2.3}$$

If $r(\mathbf{A}) = \min(m, n)$ then $\mathbf{A}$ is said to be of *full rank*; if $r(\mathbf{A}) < \min(m, n)$ then $\mathbf{A}$ is of *deficient rank*. For example, consider

$$\mathbf{A} = \begin{pmatrix} 1 & 2 & 2 \\ 1 & 3 & 0 \\ 1 & 2 & 1 \end{pmatrix} \tag{2.4}$$

This matrix is of full rank 3, since no column can be expressed as a linear combination of other columns. However, if

$$\mathbf{A} = \begin{pmatrix} 1 & 2 & 1 \\ 1 & 3 & 2 \\ 1 & 2 & 1 \end{pmatrix} \tag{2.5}$$

then $r(\mathbf{A})=2$ since the third column is exactly the difference of columns one and two. Now consider the rectangular matrix $\mathbf{B}$ given by

$$\mathbf{B} = \begin{pmatrix} 1 & 2 & 2 \\ 1 & 3 & 0 \end{pmatrix} \tag{2.6}$$

The rank of $\mathbf{B}$ cannot exceed 2; the third column is necessarily linearly dependent upon the first two since we can always find values x and y to satisfy

$$x + 2y = 2 \tag{2.7}$$
$$x - 3y = 0$$

The matrix is of full rank 2, as long as the second column (row) is not a linear function of the first.

The rank of a matrix is an indication of how much non-redundant information the matrix contains. The usual multivariate data matrix $\mathbf{X}$, introduced in Chapter 1, is generally of full rank unless some of the variables are linear combinations of others; for example, if subset scores and a total test are included, or if scores are percentages that add up to 100 for each subject.

The condition of full rank is a prerequisite for the inversion of a square matrix, and also for an operation known as *factoring*, which is important in principal components analysis (see Chapter 4), and factor analysis (see Chapter 13).

2.2.2   Eigenvalues and Eigenvectors

If **A** is a square matrix, **x** is a column vector and λ is a scalar quantity such that

$$\mathbf{A}\mathbf{x} = \lambda\mathbf{x} \tag{2.8}$$

then **x** is said to be an *eigenvector* of **A** and λ is said to be an *eigenvalue*.

Eigenvalues and eigenvectors arise frequently in statistics, particularly in situations where one wishes to find a linear combination of variables that has maximum variance, see for example, Chapter 4. Finding the eigenvalues and eigenvectors of a matrix involves laborious calculations but fortunately many very efficient computer programs are now available. In the case of a 2 x 2 matrix, however, they are relatively easy to obtain as we shall demonstrate using the matrix

$$\mathbf{A} = \begin{pmatrix} 6 & 3 \\ 3 & 4 \end{pmatrix} \tag{2.9}$$

First it can be shown (see Maxwell 1977, Chapter 3 for details) that the eigenvalues are obtained as roots of the determinantal equation

$$|\mathbf{A} - \lambda\mathbf{I}| = 0 \tag{2.10}$$

where **I** is the identity matrix of appropriate order. Expanding the left-hand side of equation (2.10) leads to the following quadratic equation

$$\lambda^2 - 10\lambda + 15 = 0 \tag{2.11}$$

the roots of which are $\lambda_1 = 8.162$ and $\lambda_2 = 1.838$. To find the eigenvector corresponding to λ_1 we use (2.8) to give

$$-2.162x_1 + 3x_2 = 0 \tag{2.12}$$
$$-3x_1 - 4.162x_2 = 0$$

It is easy to see from equation (2.8) that the elements of an eigenvector are only determined up to multiplication by a scalar; consequently equations (2.12) only define the ratio of x_1 to x_2, giving

$$\frac{x_1}{x_2} = 1.39 \tag{2.13}$$

Choosing $x_1 = 1.0$ gives $x_2 = 0.721$. Determination of the second eigenvector is left as an exercise for the reader.

Let us suppose we have a $(p \times p)$ symmetric matrix Σ with elements σ_{ij} eigenvalues, $\lambda_1, \ldots, \lambda_p$, and corresponding eigenvectors $\mathbf{c}_1, \ldots, \mathbf{c}_p$. Some useful results are as follows:

(a)

$$\sum_{i=1}^{p} \sigma_{ii} = \sum_{i=i}^{p} \lambda_i \tag{2.14}$$

(This sum is usually written as trace (Σ).)

(b)   The determinant of Σ is given by

$$|\Sigma| = \prod_{i=i}^{p} \lambda_i \tag{2.15}$$

(c) For two unequal eigenvalues the corresponding eigenvectors are *orthogonal*, i.e. if $\mathbf{c}_1$ and $\mathbf{c}_2$ are the two eigenvectors, then

$$\mathbf{c}_1'\mathbf{c}_2 = 0 \qquad (2.16)$$

(d) If the eigenvalues of Σ are distinct and we form a $(p \times p)$ matrix, $\mathbf{C}$, whose *ith* column is the *normalised* eigenvector $\mathbf{c}_i$, i.e. the eigenvector with its elements scaled so that

$$\mathbf{c}_i'\mathbf{c}_i = 1 \qquad (2.17)$$

then

$$\mathbf{C}'\mathbf{C} = \mathbf{I} \qquad (2.18)$$

and

$$\mathbf{C}'\Sigma\mathbf{C} = \Lambda \qquad (2.19)$$

where Λ is a diagonal matrix whose diagonal elements are $\lambda_1 \cdots \lambda_p$.(Any matrix satisfying equation (2.18) is known as an *orthogonal* matrix.)

2.2.3   Quadratic Forms

A *quadratic form* in p variables $x_1, \ldots, x_p$ is a homogeneous function consisting of all possible second order terms, namely

$$a_{11}x_1^2 + \cdots + a_{pp}x_p^2 + a_{12}x_1x_2 + \cdots + a_{p-1p}x_{p-1}x_p = \sum_{i,j}^{p} a_{ij}x_ix_j \qquad (2.20)$$

This expression can be conveniently written as $\mathbf{x}'\mathbf{A}\mathbf{x}$ where $\mathbf{x}' = [x_1, \ldots, x_p]$ and a_{ij} is the *ij*th element of $\mathbf{A}$. A square, symmetric matrix $\mathbf{A}$ and its associated quadratic form is called

(a) *positive definite* if $\mathbf{x}'\mathbf{A}\mathbf{x} > 0$ for every $\mathbf{x}$ not equal to the null vector, or
(b) *positive semidefinite* if $\mathbf{x}'\mathbf{A}\mathbf{x} \geq 0$ for every $\mathbf{x}$ not equal to the null vector.

Positive definite quadratic forms have matrices of full rank, the eigenvalues of which are all greater than zero. Positive semidefinite quadratic forms have matrices which are not of full rank, and if their rank is m they will have m positive eigenvalues and $p - m$ zero eigenvalues.

The matrix $\mathbf{C}$, formed from the normalised eigenvectors of $\mathbf{A}$, transforms the quadratic form of $\mathbf{A}$ to the reduced form involving only squared terms, i.e. writing $\mathbf{y} = \mathbf{C}'\mathbf{x}$ we have

$$\mathbf{C}\mathbf{y} = \mathbf{x} \quad \text{(since } \mathbf{C}\mathbf{C}' = \mathbf{I}) \qquad (2.21)$$

and

$$\mathbf{x}'\mathbf{A}\mathbf{x} = \mathbf{y}'\mathbf{C}'\mathbf{A}\mathbf{C}\mathbf{y} \qquad (2.22)$$
$$= \mathbf{y}'\Lambda\mathbf{y} \qquad (2.23)$$
$$= \lambda_1 y_1^2 + \cdots + \lambda_p y_p^2 \qquad (2.24)$$

These results are particularly relevant to principal components analysis (see Chapter 4).

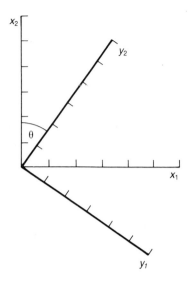

Figure 2.1 Rotation of axis

2.3 Rotation of Axes

Rotation of axes is a common procedure in principal components analysis (see Chapter 4) and factor analysis (see Chapter 13). If a point P has coordinates (x_1, x_2) with reference to a given set of orthogonal axes and if the axes are rotated about the origin through an angle θ in a clockwise direction, the coordinates (y_1, y_2) of P with reference to the rotated axes are obtained by multiplying the row vector $\mathbf{x}' = (x_1, x_2)$ by an orthogonal matrix $\mathbf{U}$ where

$$\mathbf{U} = \begin{pmatrix} \cos\theta & \sin\theta \\ -\sin\theta & \cos\theta \end{pmatrix} \tag{2.25}$$

i.e.

$$\mathbf{y}' = \mathbf{x}'\mathbf{U} \tag{2.26}$$

where $\mathbf{y}' = (y_1, y_2)$. Such a rotation is illustrated in Figure 2.1. This rotation preserves Euclidean distances of points from the origin (those readers unfamiliar with the formula for Euclidean distance should refer to page 67).

2.4 Some Basic Topics in Multivariate Statistics

In the univariate case it is often necessary to summarise a data set by calculating its mean and variance. To summarise multivariate data sets we need to find the mean and variance of each of the p variables, together with a measure of the way each pair of variables is related. For the latter the covariance or correlation of each pair of variables is used. These quantities are defined below.

2.4.1   Means

The population mean vector $\boldsymbol{\mu}' = [\mu_1, \ldots, \mu_p]$, where

$$\mu_i = E(x_i) \tag{2.27}$$

An estimate of $\boldsymbol{\mu}$ based on n, p-dimensional observations is $\bar{\mathbf{x}}' = [\bar{x}_1, \ldots, \bar{x}_p]$, where $\bar{x}_i$ is the sample mean of variable x_i.

2.4.2   Variance

The vector of population variances is $\boldsymbol{\sigma}' = [\sigma_1^2, \ldots, \sigma_p^2]$ where

$$\sigma_i^2 = E(x_i^2) - \mu_i^2 \tag{2.28}$$

An estimate of $\boldsymbol{\sigma}$ based on n, p-dimensional observations is $\mathbf{s}' = [s_1^2, \ldots, s_p^2]$, where s_i^2 is the sample variance of variable x_i.

2.4.3   Covariance

The covariance of two variables x_i and x_j is defined by

$$\text{Cov}(x_i, x_j) = E(x_i x_j) - \mu_i \mu_j \tag{2.29}$$

If $i = j$, we note that the covariance of a variable with itself is simply its variance and there is therefore no need to define variances and covariances independently in the multivariate case.

The covariance of x_i and x_j is usually denoted by σ_{ij} (so the variance of variable x_i is often denoted by σ_{ii} rather than σ_i^2).

With p variables, $x_1, \ldots, x_p$, there are p variances and $\frac{1}{2}p(p-1)$ covariances. In general these quantities are arranged in the $p \times p$ symmetric matrix, $\boldsymbol{\Sigma}$, where

$$\boldsymbol{\Sigma} = \begin{pmatrix} \sigma_{11} & \sigma_{12} & \cdots & \sigma_{1p} \\ \sigma_{21} & \sigma_{22} & \cdots & \sigma_{2p} \\ \vdots & \vdots & \ddots & \vdots \\ \sigma_{p1} & \sigma_{p2} & \cdots & \sigma_{pp} \end{pmatrix} \tag{2.30}$$

(Note that $\sigma_{ij} = \sigma_{ji}$.) This matrix is generally known as the *variance–covariance* matrix or simply as the *covariance* matrix. The sample version of $\boldsymbol{\Sigma}$, usually denoted, $\mathbf{S}$, is generally estimated as

$$\mathbf{S} = \frac{1}{n-1} \sum_{i=1}^{n} (\mathbf{x}_i - \bar{\mathbf{x}})(\mathbf{x}_i - \bar{\mathbf{x}})' \tag{2.31}$$

2.4.4   Correlation

The covariance is often difficult to interpret because it depends on the units in which the two variables are measured; consequently it is often standardised by dividing by the product of the standard deviations of the two variables to give a quantity called the *correlation coefficient*, ρ_{ij}, where

$$\rho_{ij} = \frac{\sigma_{ij}}{\sqrt{\sigma_{ii}\sigma_{jj}}} \tag{2.32}$$

The correlation coefficient lies between -1 and 1 and gives a measure of the *linear* relationship of variables x_i and x_j. It is positive if high values of x_i are associated with high values of x_j and negative if high values of x_i are associated with low values of x_j.

With p variables there are $\frac{1}{2}p(p-1)$ distinct correlations which may be arranged in a $p \times p$ matrix $\mathbf{R}$ whose diagonal elements are unity. This matrix may be written in terms of the covariance matrix $\boldsymbol{\Sigma}$ as follows:

$$\mathbf{R} = \mathbf{D}^{-\frac{1}{2}}\boldsymbol{\Sigma}\mathbf{D}^{-\frac{1}{2}} \tag{2.33}$$

where

$$\mathbf{D}^{\frac{1}{2}} = \text{diag}(\sqrt{\sigma_{ii}})$$

In most situations we will be dealing with covariance and correlation matrices of full rank, p, so that both matrices will be non-singular. In some circumstances, however, variables may be so highly correlated that the covariance or correlation matrix may be 'nearly singular'. (See comments in Chapter 8 on ridge regression.)

Many of the methods of analysis to be described in this text involve *linear compounds* of the original variables, $x_1 \cdots x_p$, that is variables constructed thus:

$$y = a_1 x_1 + a_2 x_2, \ldots, + a_p x_p \tag{2.34}$$

$$y = \mathbf{a}'\mathbf{x}$$

where $\mathbf{a}' = [a_1, \ldots, a_p]$. The variable y has mean given by

$$E(y) = \mathbf{a}'E(\mathbf{x}) = \mathbf{a}'\boldsymbol{\mu} \tag{2.35}$$

and variance

$$V(y) = E[(\mathbf{a}'(\mathbf{x} - \boldsymbol{\mu})^2] \tag{2.36}$$

Now $\mathbf{a}'(\mathbf{x} - \boldsymbol{\mu})$ is a scalar, and thus is equal to its transpose, so (2.36) may be rewritten as

$$V(y) = E[\mathbf{a}'(\mathbf{x} - \boldsymbol{\mu})(\mathbf{x} - \boldsymbol{\mu})]'\mathbf{a}] \tag{2.37}$$

$$= \mathbf{a}'[E(\mathbf{x} - \boldsymbol{\mu})(\mathbf{x} - \boldsymbol{\mu})']\mathbf{a}$$

$$= \mathbf{a}'\boldsymbol{\Sigma}\mathbf{a} \tag{2.38}$$

2.5   Maximum Likelihood Estimation

Of primary concern in many analyses is the estimation of the parameters of a statistical distribution or of a statistical model. Many methods of estimation have been developed, and for a comprehensive account of this branch of statistics readers are referred to Kendall and Stuart (1980). In this section we shall concentrate on introducing the basic ideas of *maximum likelihood* estimation, which is perhaps the most widely used of these techniques. To introduce this method let us assume that we have a single random variable, x, whose density function is $f(x; \theta)$ where θ is the parameter to be estimated. For a sample of n observations of this variable $x_1, \ldots, x_n$, the *likelihood function*, L, is defined by

$$L = f(x_1; \theta)f(x_2; \theta) \cdots f(x_n; \theta) \tag{2.39}$$

$$= \prod_{i=1}^{n} f(x_i; \theta)$$

The maximum likelihood estimate, $\hat{\theta}$, is defined as that value of θ which maximises L. It may be found by solving the equation

$$\frac{dL}{d\theta} = 0 \qquad (2.40)$$

In many cases it may be more convenient to maximise $L = \ln L$; this involves solving the equation

$$\frac{dL}{d\theta} = 0 \qquad (2.41)$$

Both equations (2.40) and (2.41) will lead to the same value for $\hat{\theta}$. To illustrate the method let us consider a random variable with an exponential distribution, that is

$$f(x;\theta) = \theta e^{-\theta x} \qquad (2.42)$$

The likelihood function for the n observations $x_1, \ldots, x_n$ takes the form

$$L = \theta e^{-\theta x_1} \theta e^{-\theta x_2} \cdots \theta e^{-\theta x_n} \qquad (2.43)$$

$$= \theta n \exp -(\theta \sum_{i=1}^{n} x_i)$$

Taking logs gives

$$L = n \ln \theta - \theta \sum_{i=1}^{n} x_i \qquad (2.44)$$

Differentiating L with respect to θ gives

$$\frac{d L}{d\theta} = \frac{n}{\theta} - \sum_{i=1}^{n} x_i \qquad (2.45)$$

Setting equation (2.45) to zero leads to

$$\hat{\theta} = \frac{n}{\sum_{i=1}^{n} x_i} \qquad (2.46)$$

In some cases the distribution of x will depend upon more than a single parameter, that is θ will be a vector rather than a scalar. Perhaps the best-known example is the normal distribution

$$f(x;\mu,\sigma^2) = \frac{1}{\sqrt{2\pi}\sigma} \exp -\frac{1}{2}\left(\frac{x-\mu}{\sigma}\right)^2 \qquad (2.47)$$

For this distribution the log-likelihood function, L, is given by

$$L = -\frac{1}{2\sigma^2} \sum_{i=1}^{n} (x_i - \mu)^2 - \frac{1}{2} n \ln \sigma^2 - \frac{1}{2} n \ln 2\pi \qquad (2.48)$$

So that

$$\frac{\partial L}{\partial \mu} = \frac{1}{\sigma^2} \sum_{i=1}^{n} (x_i - \mu) \qquad (2.49)$$

$$\frac{\partial L}{\partial \sigma^2} = \frac{1}{2\sigma^4} \sum_{i=1}^{n} (x_i - \mu)^2 - \frac{n}{2\sigma^2} \qquad (2.50)$$

Setting equations (2.49) and (2.50) to zero leads to the estimates

$$\hat{\mu} = \frac{1}{n} \sum_{i=1}^{n} x_i \qquad (2.51)$$

$$\hat{\sigma}^2 = \frac{1}{n} \sum_{i=1}^{n} (x_i - \bar{x})^2 \qquad (2.52)$$

Maximum likelihood estimates (MLE) have a number of desirable properties which are also discussed in Kendall and Stuart (1980). One important result which should be noted is that the variances and covariances of the estimates are obtained from the inverse of the information matrix, $\mathbf{I}$ which has elements given by

$$i_{jj} = E(\frac{\partial \ln f}{\partial \theta_j})^2, \qquad (2.53)$$

and

$$i_{jk} = E(\frac{\partial \ln f}{\partial \theta_j} \cdot \frac{\partial \ln f}{\partial \theta_k}). \qquad (2.54)$$

The covariance matrix of the maximum likelihood estimators is given by $(n\mathbf{I})^{-1}$ where n is the sample size. This may be estimated by substituting the MLE themselves for the parameter values in $\mathbf{I}$ (see Exercise 2.4). The information matrix is important for a number of reasons (see Kendall and Stuart, 1980).

In the simple examples given above, the equations defining the MLE (for example, equations (2.49) and (2.50) may be solved explicitly. In some cases however this is not so, since they involve complex non-linear expressions in the parameters. In such cases the equations must be solved by some type of iterative procedure, of which the most well known is Newton's method (again, see Kendall and Stuart, 1980). Alternatively the maximisation of L or $\mathcal{L}$ may be attacked directly using the alternative optimisation techniques that are described in the next section.

2.6 Optimisation Methods

Optimisation in its simplest form is concerned with finding the maximum or minimum value of a mathematical function. For example, Figure 2.2 shows a plot of the function $f(x) = x^2$, and we can see that this takes its minimum value at $x = 0$. From this figure we can see that if a point x^* corresponds to the minimum value of $f(x)$, the same point also corresponds to the maximum value of the negative of the function $-f(x)$. Consequently, without loss of generality, *optimisation* can be taken to mean minimisation, since the maximum of a function can be found by seeking the minimum of the negative of the same function.

There is no single method available for solving all optimisation problems efficiently, so that a variety of techniques have been developed for dealing with different types of problem. The simplest optimisation problems concern functions of a single variable, $f(x)$, and it is easy to show that a *necessary* condition for a minimum of $f(x)$ is that

$$\frac{df}{dx} = 0 \qquad (2.55)$$

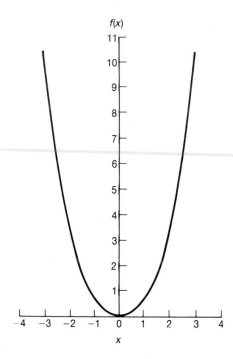

Figure 2.2   Plot of $f(x) = x^2$

A *sufficient* condition for a point satisfying these equations to be a minimum is that

$$\frac{d^2f}{dx^2} > 0 \qquad (2.56)$$

For example, suppose we wish to find the minimum value of the function, $f(x) = (x - a)^2$, where a is some constant. We have

$$\frac{df}{dx} = 2(x - a) \qquad (2.57)$$

Setting equation (2.57) to zero gives $x = a$. To check that this corresponds to a minimum we need to evaluate the second derivative of f with respect to x:

$$\frac{d^2f}{dx^2} = 2 \qquad (2.58)$$

This is always greater than zero so that the point $x = a$ corresponds to a minimum value of $f(x)$.

For a function of several variables corresponding conditions to equations (2.55) and (2.56) can be found for a minimum ; details are given in Box *et al.* (1969).

From a practical point of view the disadvantages of trying to determine the minimum of a function from consideration of equations such as (2.55) and (2.56) are considerable. A major problem is that in many cases equation (2.55) — or its equivalent in situations involving several variables — will be formidable, and its solution presents considerable difficulties. Consequently a different approach is adopted which requires an initial point x_0 to be specified and proceeds by generating a sequence of points $x_1, x_2, \ldots$ which represent improved approximations to the solution, that is

$$f(x_{i+1}) \leq f(x_i) \tag{2.59}$$

Such *iterative techniques* fall into two major classes, *direct search methods* and *gradient methods*. Here we restrict ourselves to a brief description of one technique from the latter class, namely the method of *steepest descent*; details of many other methods from both classes are available in Box *et al.* (1969).

2.6.1 Steepest Descent

Suppose we wish to find the minimum of a function of several variables $x_1, \ldots, x_m$. Starting from a given initial point, $\mathbf{x}_0' = [x_1^0, \ldots, x_m^0]$, we wish to generate a sequence of points $\mathbf{x}_1, \mathbf{x}_2, \ldots$ such that

$$f(\mathbf{x}_{i+1}) \leq f(\mathbf{x}_i) \tag{2.60}$$

The method of steepest descent generates this sequence from the following general iterative rule:

$$\mathbf{x}_{i+1} = \mathbf{x}_i + h_i \mathbf{d}_i \tag{2.61}$$

where h_i is a scalar known as *step size* and $\mathbf{d}_i$ is the search direction defined by

$$\mathbf{d}_i' = [-\frac{\partial f}{\partial x_1}, -\frac{\partial f}{\partial x_2}, \ldots, -\frac{\partial f}{\partial x_m}] \tag{2.62}$$

evaluated at the current point $\mathbf{x}_i' = [x_1^{(i)}, x_2^{(i)}, \ldots, x_m^{(i)}]$. The rationale behind this procedure is explained in detail in Box *et al.*, Chapter 4. Essentially, however, the greatest change in the function occurs when the components of the search direction are chosen to be proportional to the corresponding $\frac{\partial f}{\partial x_i}$, and for this change to be a decrease the constant of proportionality must be negative.

A basic application of the method would be as follows. Determine the gradient vector at the current point and, using a specified step length, h_i, obtain a new point by applying the general iterative rule (2.62). This procedure is then repeated using a constant step-length until a step which does not reduce the function value is taken. Now the step-length is reduced and the procedure continued from the best point. Iterations are continued until some convergence criterion is satisfied; for example until all the elements of $\mathbf{d}_i$ are almost zero. The performance of the method for a function of two variables is demonstrated in Figure 2.3. At point 5 the function value is greater than at point 4, so a reduction in the step-length is called for before restarting the search from point 4. Optimisation methods have become of great importance in modern statistics, and are used in many of the techniques discussed in this book. An introductory account of such techniques which pays particular attention to their application in statistics is given in Everitt (1987).

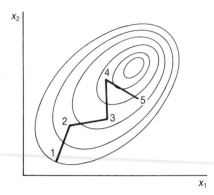

Figure 2.3 Example of steepest descent. (Reproduced with permission from Box *et al.*, 1969)

Exercises

2.1   Find the rank of the matrix

$$\mathbf{A} = \begin{pmatrix} 1 & 3 & 1 & 2 & 1 \\ 8 & 2 & 1 & 2 & 1 \\ 8 & 6 & 8 & 1 & 1 \\ 1 & 5 & 7 & 4 & 4 \\ 10 & 8 & 3 & 6 & 5 \end{pmatrix}$$

2.2   Find the eigenvalues and normalized eigenvectors of

$$\mathbf{B} = \begin{pmatrix} 2 & 1 & 1 \\ 1 & 2 & 1 \\ 1 & 1 & 2 \end{pmatrix}$$

2.3   Given that $\mathbf{A}$ is of order $(m \times n)$, and $\mathbf{B}$ and $\mathbf{C}$ are square non-singular matrices of order $(m \times m)$ and $(n \times n)$ respectively, show that

$$\text{rank}(\mathbf{A}) = \text{rank}(\mathbf{BA}) = \text{rank}(\mathbf{AC})$$

2.4   Given that $\mathbf{x}$ and $\mathbf{y}$ are vectors of the same order, prove that

$$\mathbf{x}'\mathbf{y} = \text{trace } \mathbf{yx}'$$

2.5   Find the information matrix for the maximum likelihood estimates of the two parameters of the normal distribution.

2.6   Given that $x_1, x_2, \ldots, x_n$ are independent random variables, and x_j has a binomial distribution with parameters n_j and $p(j = 1, \ldots, N)$, show that the maximum likelihood estimator of p is

$$\hat{p} = \frac{\sum_{j=1}^{N} x_j}{\sum_{j=1}^{N} n_j}$$

2.7 Given that $\mu'_r = E(x^r)$ and $\mu_r = E(x - E(x))^r$, show that

$$\mu_2 = \mu'_2 - \mu'^2_1$$
$$\mu_4 = \mu'_4 - 4\mu'_3\mu'_1 - 6\mu'_2\mu'^2_1 - 3\mu'^4_1$$

2.8 Show that the function $f(x) = x^2 + 2x + 3$ has a minimum at $x = -1$.

2.9 Find the minimum of the function

$$(x_1 - x_2 + x_3)^2 + (x_1 + x_2 - x_3)^2$$

2.10 Use the method of steepest descent to find the minimum of the function in Exercise 2.8, starting from the point $x = 3$ with step size $h = 0.2$.

2.11 Given that the random vector $x' = [x_1, x_2, x_3]$ has covariance matrix Σ given by

$$\Sigma = \begin{pmatrix} \sigma_{11} & \sigma_{12} & \sigma_{13} \\ \sigma_{21} & \sigma_{22} & \sigma_{23} \\ \sigma_{31} & \sigma_{32} & \sigma_{33} \end{pmatrix}$$

find the covariances of the variables y_1 and y_2 given by

$$y_1 = x_1 + x_2 + x_3$$
$$y_2 = x_1 - x_3$$

PART II
EXPLORING MULTIVARIATE DATA

Chapters 3 to 6 cover a variety of methods which are primarily of use for the exploration of multivariate data. In a very general sense their aim might be considered to be that of detecting patterns or indicating potentially interesting relationships in the data. Only when some pattern is thought to exist can the further steps be taken of setting up models and hypotheses for future investigation. (In many cases, of course, researchers might feel that an informative description of their data provided by one or other of the techniques to be discussed in this section completes the analysis.)

A theme common to many of the methods to be described in this part of the test is the production of results in the form of a graph or some other type of visual display. This often allows the investigator to get a 'view' of the data that might have an impact impossible or at least difficult to achieve by the mere examination of statistics derived from the data.

3

The Initial Examination of Multivariate Data

3.1 Introduction

Most statistical analysis of data is, nowadays, carried out with the aid of computers, particularly through the use of the type of statistical package described in Appendix A. This is obviously to be welcomed because it frees the investigator from the burden of large amounts of repetitive arithmetic. Indeed many analyses now performed routinely would simply not be possible without the aid of a computer. However, the wide availability of packages with which complex analyses such as *multidimensional scaling* (Chapter 5), *multiple regression* (Chapter 8) or *factor analysis* (Chapter 13) can be carried out relatively easily, is not without its dangers. One of these has been to foster in some research workers the unfortunate tendency to rush into the use of complicated methods of analysis without first considering some simple preliminary techniques designed to give them a general 'feel' of the data. Such techniques may serve several purposes. They may for example give an early indication of the presence of *outliers* (that is observations which appear to be inconsistent with the rest of the data). These may be caused by recording errors, or by errors introduced when transferring the data to a computer file, in which case they may be corrected and then included in future analyses. On the other hand outliers may be genuine observations indicative perhaps of an unusual distribution of the data. In either case outliers need to be detected since they can have the effect of greatly distorting the results of analyses carried out by many of the techniques to be discussed in later chapters. (A detailed account of the detection of outliers is given in Barnett and Lewis, 1978.)

In the early phase of an analysis the techniques to be discussed in this chapter might prove helpful in indicating which of the more complicated exploratory techniques could usefully be applied to the data. For example, an Andrews plot (see Section 3.3) might suggest the presence of distinct 'clusters' of observations, indicating that further investigation via some form of multidimensional scaling or cluster analysis might be fruitful.

3.2   Marginal Views of Multivariate Data

Since multivariate data consists of a number of variables recorded for each of the individuals in the study, traditional descriptive statistics and their modern counterparts could be used to examine one or two variables at a time. Histograms, for example, could be used to look at the distribution of values on each variable, and simple scatterplots used to obtain a number of two-dimensional views of the data. Similarly methods from Tukey's collection of exploratory data techniques (see Tukey, 1977), such as *stem-and-leaf* and *box plots* might be used. Additionally Chatfield's IDA *initial data analysis* approach might be considered (see Chatfield, 1985, 1988).

A further use of one and two-dimensional *marginal news* of multivariate data is for suggesting *transformations* of variables which might be helpful in later analyses. A transformation of a set of variable values $x_1, x_2, \ldots, x_n$ is a function T that replaces each x_i by a new value $T(x_i)$ so that the transformed variables values are $T(x_1), T(x_2), \ldots, T(x_n)$. The transformed values may overcome some of the distributional problems of the raw data such as (1) strong asymmetry, (2) many outliers in one tail etc., which can adversely affect some types of analysis. (A detailed discussion of transforming data is given in Emerson and Stoto, 1983.)

As an example of looking at marginal views of a set of multivariate data consider Table 3.1, which consists of observations on *IQ, birth weight of child,* and *age* of 25 mothers who had just had their first child. Figure 3.1 shows the stem-and-leaf plot of each variable, and Figure 3.2 a scattergram for each pair of variables.

The stem-and-leaf plots show nothing very disturbing about the distributions of the three variables, although the maximum value of the weight variable is a little removed from the remainder of the observations. Otherwise the distributions are symmetrical, and there seems to be little reason for considering any transformations.

The two scatterplots involving weight do however give rather more evidence of some 'unusual' observations which might perhaps be regarded as outliers. These observations are circled. The correlations between age and weight and between IQ and weight, when all observations are included, are -0.50 ($p = 0.012$) and -0.11 ($p = 0.59$) respectively. If the suspect observations are removed the corresponding correlations are -0.15 ($p = 0.49$) and 0.40 ($p = 0.06$). Clearly the two marked observations are having a pronounced effect and consideration would need to be given to how to handle them in subsequent analyses.

Now consider the data in Table 3.2 which gives the salaries as of 1979, quoted in Swiss francs, of seven different professions in 44 cities around the world. The boxplots for the standardised salaries of each profession are shown in Figure 3.3. The distributions are very similar and largely symmetrical. Scatterplots of the salaries for each pair of professions are shown in Figure 3.4. Arranging such scatterplots in this lower triangular form provides an overall impression of the relationships which is often very useful. Here for example it is easy to see that the strength of the relationship between the various pairs differs considerably. The salaries of teachers appear to be strongly related to those of all other professions. We shall return to these data later in this chapter.

Table 3.1   Birth weight data

IQ	Weight (g)	Age
125	2536	28
86	2505	31
119	2652	32
113	2573	20
101	2382	30
143	2443	30
132	2617	27
106	2556	36
121	2489	34
109	2415	29
88	2434	27
116	2491	24
102	2345	26
75	2350	23
90	2536	24
109	2577	22
104	2464	35
110	2571	24
96	2550	24
101	2437	23
95	2472	36
117	2580	21
115	2436	39
138	2200	41
85	2851	17

(a) *IQ*

7:5
8:568
9:056
10:1124699
11:035679
12:15
13:28
14:3

(b) *Weight*

22:0
23:458
24:134446799
25:044567788
26:25
28:51

(c) *Age*

1:7
2:012334444
2:67789
3:00124
3:5669
4:1

Figure 3.1   Stem-and-leaf plots for age, birth weight and IQ data

Although marginal views of multivariate data are useful they do not always give the complete picture, since they do not always reflect adequately the *multidimensional structure* of the data. Figure 3.5, for example, illustrates a situation where the univariate, marginal distributions do not indicate the complex structure of the original bivariate data. Additionally if many variables are involved the sheer number of one and two-dimensional views may become extremely confusing.

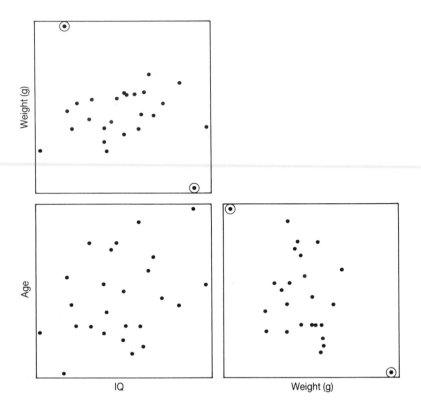

Figure 3.2   Scatterplots for age, birth weight and IQ data

Clearly it would be an advantage if simple plots of multivariate data could be obtained which contained information from *more* than one or two variables, and a number of ingenious suggestions have been made for this purpose. For example, Gower (1967) reports a method due to Ross in which two variables are used to derive a scatterplot of the data. Information from a third variable is now included in this diagram by drawing, from each point, a line with length proportional to that observation's value of this third variable and directed eastwards if this value is positive and westward if negative. Similarly for a fourth variable the north/south direction could be used and for variables five and six the NE/SW and NW/SE directions. An example of this type of plot is given in Everitt (1978).

In the same vein is a technique known as a *weathervane plot*, introduced by Bruntz *et al.* (1974). The authors were interested in analysing air pollution data, and some of the variables of interest were amounts of solar radiation, wind speed and direction, and temperature. Figure 3.6 taken from Gnanadesikan (1977) shows a weathervane plot for a particular data set. Here the abscissa is used for the total solar radiation from 8 a.m to noon, while the ordinate is the average ozone level observed from 1 p.m. to 3 p.m. The plot is for

Table 3.2 Salaries data

	City	Teacher	Chaufr	Mechnc	Cook	Managr	Enginr	Cashr
1	Abu Dhabi	16 800	15 336	8 996	31 562	26 520	31 635	11 960
2	Amsterdam	34 125	29 820	26 642	37 067	59 280	47 730	32 200
3	Athens	11 025	10 650	12 456	19 451	31 980	18 870	16 100
4	Bahrain	10 500	5 112	6 574	14 680	56 940	44 955	18 400
5	Bangkok	3 150	3 408	3 460	12 478	14 820	7 770	3 680
6	Bogota	4 725	3 408	3 806	12 478	14 040	14 430	3 220
7	Brussels	28 350	26 412	25 258	24 589	59 280	33 855	38 640
8	Buenos Aires	5 775	6 390	6 574	36 333	21 060	36 075	13 800
9	Caracas	11 550	14 058	20 068	35 232	45 240	42 180	17 940
10	Chicago	33 600	36 636	39 790	30 094	60 060	48 285	24 380
11	Copenhagen	32 550	31 950	34 946	46 976	67 080	53 280	33 120
12	Dublin	18 375	13 206	13 840	20 919	23 400	25 530	17 020
13	Dusseldorf	33 600	33 228	23 528	31 562	63 180	44 400	30 360
14	Geneva	56 700	44 304	37 022	31 929	71 760	53 835	45 080
15	Helsinki	19 950	19 596	17 646	16 515	48 360	33 855	17 940
16	Hongkong	11 550	7 668	5 882	14 680	20 280	17 205	11 500
17	Istanbul	4 725	5 964	6 228	9 909	13 260	12 210	5 980
18	Jakarta	2 625	2 130	2 422	5 505	8 580	6 105	3 220
19	Jeddah	21 000	20 360	20 760	26 057	60 060	32 745	22 540
20	Johannesbg	14 700	13 206	16 609	19 818	31 200	36 630	13 800
21	London	20 475	18 318	17 646	14 680	31 200	21 090	17 020
22	Los Angeles	32 550	29 394	36 330	32 296	59 280	46 065	17 940
23	Luxembourg	42 000	35 784	20 068	24 956	63 960	63 270	38 640
24	Madrid	14 700	14 058	12 110	18 717	32 760	31 635	24 380
25	Manila	2 100	2 982	1 730	4 771	20 280	4 440	4 140
26	Mexico City	0 825	0 816	8 304	27 892	28 860	22 200	11 040
27	Milan	12 600	14 910	13 494	16 148	17 160	31 080	23 920
28	Montreal	29 400	25 560	23 528	20 185	51 480	34 410	15 180
29	New York	27 300	26 838	32 870	48 444	67 080	53 280	20 240
30	Oslo	25 200	25 560	25 258	31 195	42 900	42 735	28 060
31	Panama	4 725	4 260	7 266	12 478	22 620	24 420	10 120
32	Paris	24 150	24 282	15 916	30 828	40 560	43 845	21 160
33	Rio de Jan	7 350	7 242	8 650	15 047	53 040	42 735	9 660
34	S. Francisco	32 025	30 246	34 946	50 279	65 520	46 065	16 560
35	Sao Paulo	9 450	8 520	11 072	12 845	64 740	29 970	11 040
36	Singapore	8 925	4 260	5 190	10 276	24 960	8 325	9 200
37	Stockholm	28 875	24 708	25 950	33 397	54 600	33 855	26 680
38	Sydney	28 350	19 596	20 068	21 286	34 320	31 080	19 780
39	Teheran	12 600	11 076	13 840	24 956	26 520	37 185	18 400
40	Tel-Aviv	7 875	11 928	9 688	9 542	14 040	14 309	6 440
41	Tokyo	30 450	26 412	16 954	25 690	63 180	34 410	40 940
42	Toronto	29 925	25 986	25 950	20 919	44 460	39 960	14 260
43	Vienna	19 425	23 430	19 722	18 717	42 900	38 850	26 220
44	Zurich	52 500	42 600	34 600	36 700	7 800	55 500	46 000

a specific site, and the centres of the circles correspond to different days. These points provide information on the relationship between ozone levels and solar radiation. The diameter of the circle surrounding each such point is proportional to the observed daily maximum temperature, while the line projecting from the circle is such that its length is inversely proportional to an average wind speed. Also, if the lines are considered as arrows whose heads

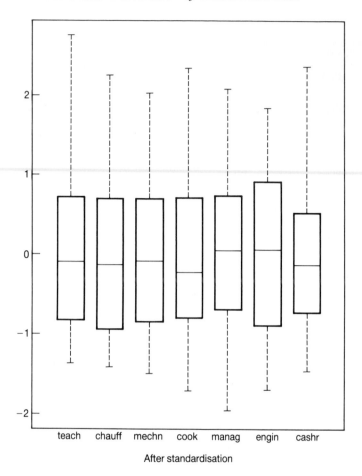

After standardisation

Figure 3.3 Boxplots for salaries data

are at the centres of the circles, then the orientations of these lines correspond to average wind directions.

Figure 3.6 displays the relationships between five variables and enables some overall patterns in the data to be detected. For example, at a given level of solar radiation, ozone seems to increase as temperature increases and wind speed decreases. Wind direction does not appear to have a major influence in these data.

Such plots will clearly become confusing if more than four or five variables are included. Consequently, their use on the raw data is likely to be limited. However, they may prove useful when applied to a reduced number of dimensions obtained from a technique such as principal components analysis (see Chapter 4). (Some other ideas for including extra information on scatterplots are given in Tukey and Tukey, 1981c.)

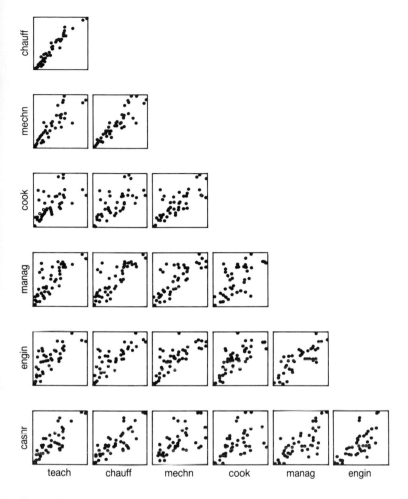

Figure 3.4 Scatterplots for salaries data

A further development in the use of marginal news for the exploratory analysis of multivariate data is that described by Friedman and Tukey (1974) and known as *projection pursuit*. This technique seeks out linear projections of the multivariate data into a line or a plane (principal component plots in two dimensions represent one example of the latter; see Chapter 4). Selecting from the large number of possible projections for high-dimensional data is achieved by the local optimisation over projection directions of some index of 'interestingness'. Friedman and Tukey for example suggested an index specifically designed to allow projections to reveal clustering in the data. Jones and Sibson (1987) give a detailed discussion of a number of possible indices and a general discussion of the whole area of projection pursuit. A number of the examples they discuss illustrate that this approach may often give more

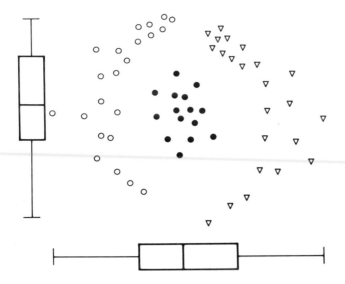

Figure 3.5   Two-dimensional multimodal data having unimodal marginals (as indicated by the box plots)

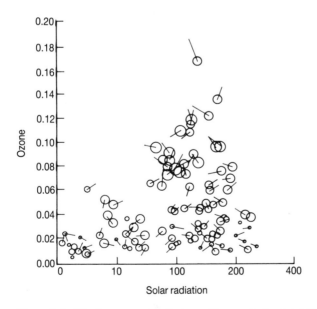

Figure 3.6   Weathervane plot. (Taken with permission from Gnanadesikan, 1977)

informative views of the data than a straightforward principal components plot.

3.3   Andrews Plots and Chernoff Faces

In this section we describe two novel techniques which give a graphical display of multivariate data and often allow an informative initial examination. Both were introduced in the 1970s; the first, due to Andrews (1972), represents each multidimensional observation as a function of a particular form. The second, due to Chernoff (1973), represents the multivariate observations as a set of cartoon faces.

3.3.1   Andrews Plots

The procedure is essentially very simple; each of the p-dimensional observations $\mathbf{x}' = [x_1, x_2, \ldots, x_p]$, is mapped into a function of the form

$$x(t) = x_1/\sqrt{(2)} + x_2 \sin(t) + x_3 \cos(t) + x_4 \sin(2t) + x_5 \cos(2t) + \cdots \qquad (3.1)$$

This function is now plotted over the range $-\pi \le t \le \pi$. A set of multivariate observations will now appear as a set of lines on the plot. The usefulness of this particular representation lies primarily in the fact that this function preserves Euclidean distances, in the sense that observations close together in the original p-dimensional space will correspond to lines on the plot that remain close together for all values of t: points far apart in the original space will be represented by lines which remain apart for at least some values of t. This property allows the plots to be examined for distinct groups of observations, outlying observations and so on. For example Figure 3.7 shows an Andrews plot for 30 five-dimensional observations. This clearly indicates the presence in the data of three well-separated groups of observations. Such a result may lead to the application of some method of cluster analysis (see Chapter 6) to these data in an effort to confirm the presence of the distinct groups, although for this particular data set, the result from the Andrews plot is so clear-cut that further, more complex analysis is really not required.

A problem which arises when using this technique is that only a fairly limited number of observations may be plotted on the same diagram before it becomes too confused to be helpful. Various procedures might be adopted in order to overcome this problem. For example, first a plot of all observations could be produced to assess the general characteristics of the sample. This could be followed by separate plots of each set of, say, ten observations which could be examined and compared to the whole. Alternatively, selected quantities or percentage points of the distribution of the n values of f could be plotted along with the curves of selected individual observations. An example of such a plot is given in Gnanadesikan (1977).

Examining the form of the function involved in Andrews plots (see equation 3.1), it is clear that the original variables are not equally weighted. Some are associated with cyclic components having a high frequency, others with components having a low frequency. Since in these plots low-frequency components are more informative than those with high frequencies, it may be useful to associate x_1 with the variable considered, in some sense, to be the most important, x_2 with the second most important, and so on. In the absence

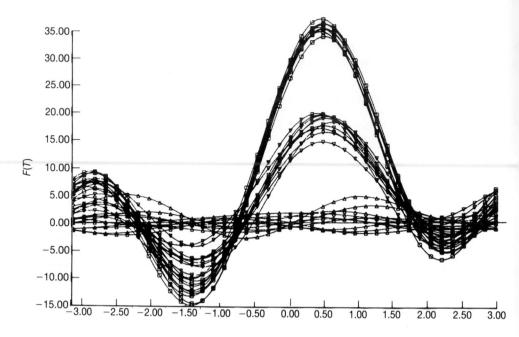

Figure 3.7   Andrews plot for 30 five-dimensional observations

of any firm ideas as to such an order of variables, it may be useful to apply Andrews' technique not to the raw data but to the transformed variables obtained from principal components analysis (see Chapter 4), since these will automatically be in order of decreasing importance in a particular sense.

Some other possible functions which could be used in place of that in equation (3.1) are given in Andrews (1972) and Gnanadesikan (1977) and an interesting application of the procedure is described by Morgan (1981).

3.3.2   Chernoff Faces

Chernoff (1973) describes a method for representing multivariate data in which each multivariate observation has a corresponding face, the features of which are governed by the values taken by particular variables. A sample of multivariate observations is now represented by a collection of such faces and these may then be examined to assess the similarities and differences between observations. For example, Figure 3.8 shows a 'faces' representation of the salary data in Table 3.2. The correspondence between face features and each variable is as follows:

(1)   Teacher: area of face.
(2)   Chauffer: shape of face.
(3)   Mechanic: length of nose.
(4)   Cook: location of mouth.

(5)   Manager: curve of smile.
(6)   Engineer: width of mouth.
(7)   Cashier: separation of eyes.

Here we shall only comment on the obvious satisfaction of the workers in Geneva and Zurich as indicated by their 'fat-smiling' faces representation. Extracting further information from the diagram is left as an exercise for the reader.

This technique has been criticised by a number of statisticians (for example, Chatfield and Collins, 1980) on the grounds of its subjectivity, since different observers are very likely to weigh facial features in different ways. Without doubt this is a problem. However, a degree of subjectivity is unlikely to be entirely absent from other graphical techniques, and a set of faces does have the advantage of providing a more interesting representation than many other techniques. Consequently investigators may be willing to spend more time and effort in studying faces and this could result in greater insights into the data. Certainly, used in particular ways, a face representation can provide a dramatic display of multivariate data. A recent addition to the faces techniques is the use of asymmetric faces (see Flury and Riedwyl, 1981), and a number of interesting applications are described in Flury and Riedwyl (1988).

3.4   Probability Plots

Probability plots are well known from univariate statistics where they are used to check distributional assumptions and to obtain rough estimates of distribution parameters. The basic procedure involves ordering the observations and then plotting them against the appropriate values of the assumed cumulative distribution function. (Details are given in Everitt, 1978, Chapter 4). For multivariate data such plots may be used to examine each variable separately. Alternatively the multivariate observation might be converted to a single number in some way. For example, if we were interested in assessing a data set for multivariate normality, we could convert each observation, x_i, into a *generalised distance* d_i^2 giving a measure of the separation of the particular observation from the mean vector of the complete sample, $\bar{x}$; d_i^2 is given by

$$d_i^2 = (\mathbf{x}_i - \bar{\mathbf{x}})'\mathbf{S}^{-1}(\mathbf{x}_i - \bar{\mathbf{x}}) \tag{3.2}$$

where $\mathbf{S}$ is the sample covariance matrix. If the data do arise from a multivariate normal distribution, then these distances have, approximately, a chi-square distribution with p degrees of freedom. So, plotting the ordered distances against the corresponding quantiles of the appropriate chi-square distribution should lead to a straight line through the origin. Figure 3.9 for example, shows such a plot for 200 obervations from a five-dimensional multivariate normal distribution.

Departure from linearity in such plots indicates that the data do not have a multivariate normal distribution. An example of this is seen in Figure 3.10, which is a chi-square plot of 200 five-dimensional observations generated from a population containing two distinct groups. Such plots are often also useful in indicating outliers as is illustrated for some five-dimensional data in Figure 3.11.

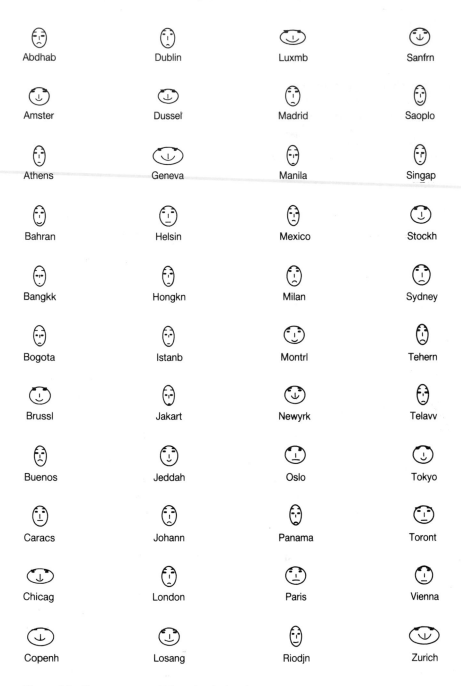

Figure 3.8   Faces representation of salaries data

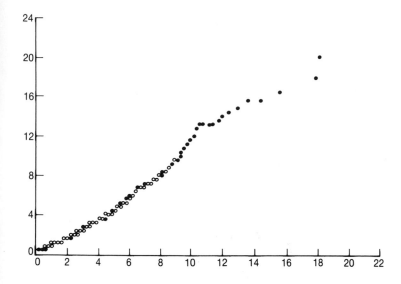

Figure 3.9 Chi-square plot of generalised distance for 200 observations from a five-dimensional multivariate normal distribution. In Figures 3.9–11 the theoretical quantiles are plotted along the *x*-axis and the ordered generalised distances along the *y*-axis

3.5 Rearranging Tables of Numerical Data

Many (if not most) research workers are happier when examining representations of their data in the form of graphs and diagrams, possibly because, as suggested by Mahon (1987), such graphs are better for indicating *relationships* than the original table of numerical data. Indeed most investigators spend little or no time examining such tables directly, preferring instead to move to some form of analysis considered appropriate. Despite this it may be worth spending a little time considering whether straightforward tables of numerical data could perhaps be made useful in some situations.

Ehrenberg (1977) argues that many tables appear to be uninformative simply because they are badly presented, and, with the application of some intuitively simple rules, many could be made far more useful. These rules consist of rounding the numbers to two significant or effective digits, placing those numbers that need to be compared in columns rather than rows, and organising the spacing and layout of the table so that the eye does not have too far to travel in making comparisons.

Table 3.3 for example, gives some data concerning unemployment in Great Britain over four selected years, as originally presented in a government publication. Rounding the data, and transposing rows and columns, gives Table 3.4. Here it is far easier to detect variations and sub-patterns. Contrary to the total trend, for example, the female figures levelled off only for 1968 and 1970. It is also clear that the 1973 figure is particularly high. When compared with Table 3.3, Table 3.4 gives a much better 'feel' for the data.

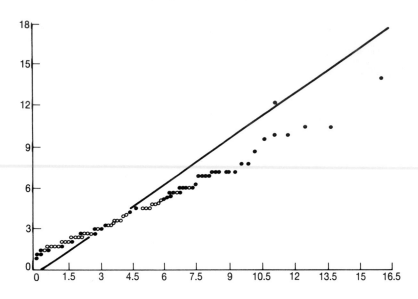

Figure 3.10   Chi-square plot for five-dimensional data containing two distinct groups

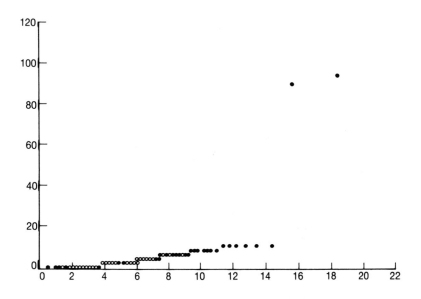

Figure 3.11   Chi-square plot indicating the presence of outliers

Rounding and a rearrangement of rows and columns can also be useful in dealing with correlation matrices. For example, Table 3.5 (reproduced from

Table 3.3 Unemployment in Great Britain—original version (Taken with permission from Ehrenberg, 1977)

	1966	*1968*	*1970*	*1973*
Total unemployed (Thousands)	330·9	549·4	582·2	597·9
Males	259·6	460·7	495·3	494·4
Females	71·3	88·8	86·9	98·5

Table 3.4 Unemployment in Great Britain—rearranged table

Year	Unemployed (000s)		
	Total	*Male*	*Female*
1966	330	260	71
1968	550	460	89
1970	580	500	87
1973	600	500	99
Average	520	430	86

Table 3.5 Correlation matrix for nine rating scales

	1	*2*	*3*	*4*	*5*	*6*	*7*	*8*	*9*
1	1·000								
2	0·638	1·000							
3	0·017	0·193	1·000						
4	0·139	0·097	0·616	1·000					
5	0·150	0·245	−0·181	−0·372	1·000				
6	0·704	0·570	0·015	0·179	0·091	1·000			
7	0·114	0·238	0·543	0·476	0·012	0·102	1·000		
8	0·160	0·196	0·302	0·100	0·252	0·007	0·213	1·000	
9	0·228	0·269	0·041	−0·052	0·533	0·172	0·145	0·447	1·000

Maxwell, 1977) shows a correlation matrix for nine ratings scales obtained from a sample of 380 adults. In the case of a relatively small matrix it is often possible, by visual examination of the elements, to discover subsets of variables which correlate relatively highly with each other. For the matrix in Table 3.5 this appears to be the case, since a close examination shows that the variates tend to fall into three groups, namely (1, 2, 6), (3, 4, 7) and (5, 8, 9). Rearranging the rows and columns according to these groups and rounding the figures gives Table 3.6, where the three-group pattern is clearly visible. This 'analysis' required nothing more than a little careful thought and a pencil and paper. Nevertheless it makes almost redundant (for this particular data set) the more formal methods of looking at the structure or pattern in correlation matrices (factor analysis, for example). Of course, for larger examples this may no longer be the case.

Table 3.6   Correlations for rating scales in rearranged order and rounded to one decimal places

	1	2	6	3	4	7	5		9
1	1·0								
2	0·6	1·0							
6	0·7	0·6	1·0						
3	0·0	0·2	0·0	1·0					
4	0·1	0·1	0·2	0·6	1·0				
7	0·1	0·2	0·1	0·5	0·5	1·0			
5	0·1	0·2	0·1	−0·2	−0·4	0·0	1·0		
8	0·2	0·2	0·0	0·3	0·1	0·2	0·4	1·0	
9	0·2	0·3	0·2	0·0	0·0	0·1	0·5	0·4	1·0

3.6   Missing Values in Multivariate Data

Table 1.1 in Chapter 1 shows a set of multivariate data which contains a number of *missing values*, reflecting a problem which is only too often met in practice. How best to deal with such data sets depends in part on how much missing data is involved and partly on the mechanism which gave rise to the missing values. If there are few missing values and the data are missing completely at random then any analysis should be based on those cases with a complete set of variable values. Apart from the potential loss of information in discarding incomplete cases, complete-case analysis raises no problems.

An alternative to considering only complete cases, which is less wasteful of information, is to include all cases which have values for a particular variable. This method, usually known as *available-case* analysis, uses all the available values; it does however have the considerable disadvantage that statistics such as means, variances, covariances etc., will be based on *different* numbers of subjects. In the context of multivariate analysis the biggest disadvantage of the available-case approach is that it can lead to covariance and correlation matrices which are **not** positive definite.

Little and Rubin (1987) discuss a number of other approaches to missing values in multivariate data which involve *estimating* the missing values from the values of those variables which are available.

3.7   Summary

The techniques and procedures in this chapter are primarily of use in the early stages of data analysis when the investigator is attempting to gain insights into any interesting patterns or relationships in the data which can be further explored by more complex techniques at some later stage of the analysis. The methods are generally easy to use and do not take up enormous amounts of computer time, and it would be encouraging to see a wider use of these preliminary procedures rather than an increase in the tendency simply to submit the data to a standard (and often irrelevant) procedure via a computer package. (Some very interesting examples of other methods for the preliminary investigation of multivariate data are given in Barnett, 1981, Kleiner and Hartigan, 1981, and Tukey and Tukey, 1981a, b and c.)

Table 3.7

1	2	3	4	5	6	7	8	9	10
1·000	0·409	0·332	0·270	0·483	−0·048	0·091	0·100	0·137	0·106
	1·000	0·285	0·266	0·452	0·138	0·201	−0·055	0·026	0·169
		1·000	0·323	0·360	0·221	0·183	0·094	0·221	0·411
			1·000	0·262	0·262	0·164	0·068	0·222	0·383
				1·000	0·201	0·208	0·165	0·234	0·252
					1·000	0·389	0·315	0·036	0·266
						1·000	0·371	0·127	0·472
							1·000	0·352	0·432
								1·000	0·330
									1·000

Exercises

3.1 Yule *et al.* (1969) administered 10 cognitive tests from the Wechsler series to 150 children, graded as good or poor readers. The correlation matrix for good readers on the ten tests is shown in Table 3.7. By suitable rounding and rearrangement of rows and columns, show that the tests tend to fall into two groups.

3.2 The following data are failure times, in days, of 45 transmissions from caterpillar tractors belonging to the Atkinson Construction company, South San Francisco.

```
4381 3953 2603 2320 1161 3286 6914 4007 3168 2376 7498 3923
     9460 4525 2168 1288 5085 2217 6922 218  1309 1875 1023
     1697 1038 3699 6142 4732 3330 4159 2537 3814 2157 7683
     5539 4839 6052 2420 5556 309  1295 3266 6679 1711 5931
```

Construct a sensible stem-and-leaf display of the data and calculate the median and midspread. (See Hartwig and Dearing 1979 and Erickson and Nosanchuk 1979.)

3.3 Table 3.8 shows the hours of work needed to earn the price of the same basket of goods in selected occupations for a number of European cities in 1973. Construct scatter plots for each pair of occupations and comment on your results.

3.4 Construct a set of Andrews plots for the cities in the above example. Do these suggest that the cities fall into distinct groups?

3.5 The data in Table 3.9 relate the number of crimes of different types per 100000 population in several American cities. A representation of these data in terms of cartoon faces is also shown (see Figure 3.12). Does the table of raw data or the set of faces give greater insights into any pattern in these data?

Table 3.8

City	Teachers	Bus drivers	Bank tellers	Secretaries
Athens	25	43	30	42
Brussels	19	24	15	20
Copenhagen	25	28	27	31
Dusseldorf	15	26	22	30
Geneva	11	21	18	28
Helsinki	17	26	33	34
Istanbul	36	64	46	39
Lisbon	14	42	18	22
London	18	20	18	20
Luxembourg	10	16	14	25
Madrid	13	38	24	34
Milan	22	23	13	27
Oslo	18	22	21	27
Paris	15	21	20	26
Rome	25	20	15	32
Stockholm	18	25	24	32
Vienna	19	26	22	30
Zurich	11	21	18	26

Table 3.9

City	Murder	Rape	Robbery	Assault	Burglary	Larceny	Auto theft
Atlanta	16·5	24·8	106	147	1112	905	494
Boston	4·2	13·3	122	90	983	669	954
Chicago	11·6	24·7	340	242	808	609	645
Dallas	18·1	34·2	184	293	1668	901	602
Denver	6·9	41·5	173	191	1534	1368	780
Detroit	13·0	35·7	477	220	1566	1183	788
Hartford	2·5	8·8	68	103	1017	724	468
Honolulu	3·6	12·7	42	28	1457	1102	653
Houston	16·8	26·6	289	186	1509	787	697
Kansas C.	10·8	43·2	255	226	1494	955	765
LA	9·7	51·8	286	355	1902	1386	862
New Orleans	10·3	39·7	266	283	1056	1036	776
NY	9·4	19·4	522	267	1674	1392	848
Portland	5·0	23·0	157	144	1530	1281	488
Tucson	5·1	22·9	85	148	1206	756	483
Washington	12·5	27·6	524	217	1496	1003	739

(Data taken, with permission, from Hartigan, 1975).

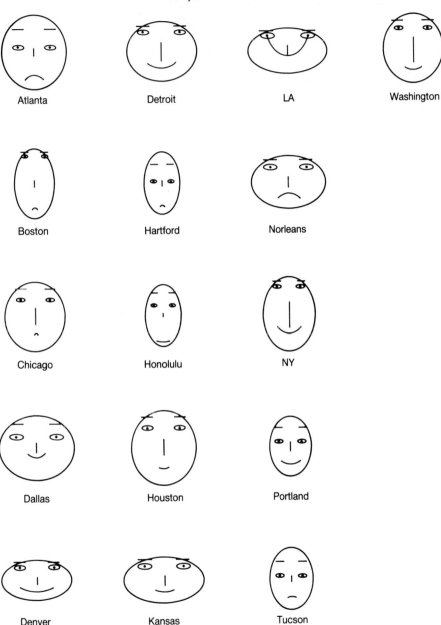

Figure 3.12 Chernoff faces for city crime data

4

Reducing the Dimensionality of Multivariate Data: Principal Components and Correspondence Analysis

4.1 Introduction

Principal components analysis is amongst the oldest and most widely used of multivariate techniques. Originally introduced by Pearson (1901) and independently by Hotelling (1933), the basic idea of the method is to describe the variation of a set of multivariate data in terms of a set of uncorrelated variables each of which is a particular linear combination of the original variables. The new variables are derived in decreasing order of importance so that, for example, the first principal component accounts for as much as possible of the variation in the original data. The usual objective of this type of analysis is to see whether the first few components account for most of the variation in the original data. It is argued that if they do, then they can be used to summarise the data with little loss of information, thus providing a reduction in the dimensionality of the data, which might be useful in simplifying later analyses.

For example, consider a situation where one has data consisting of several examination results for each of a number of students. A question of interest might be how to construct an informative index of examination achievement. One obvious index is the mean score, although if the possible or observed range of examination scores varies from one subject to another it might be more sensible to produce some kind of weighted mean, or alternatively standardise the results for the separate examinations before attempting to combine them. Another procedure would be to use the first principal component derived from the observed examination results. This would give an index providing maximum discrimination between the students, with examination scores that vary most within the sample of students being given the highest weight. A further possible application arises in the field of economics, where complex data are often summarised by some kind of index number, for example, indices of prices, wage-rates, cost of living and so on. When assessing changes in prices over time, for instance, the economist will wish to allow for the fact that prices of some commodities are more variable than others, or that the prices of some of the commodities are considered more important than others, and in each case will weight his index accordingly. In such cases the first principal component can often satisfy the investigator's requirements (see Kendall, 1975).

But often the first principal component will not give an adequate description of the multivariate data, or this component may not be the one that is the most interesting to the investigator. A taxonomist, for example, when investigating variation in morphological measurements on animals would be more interested in the second principal component since this is likely to be an indication of shape, whereas the first will probably be merely an indicator of the animal's size. Often the first principal component derived from clinical psychiatric scores on patients will provide only an index of the severity of symptoms, and it is the second or subsequent components that gives the investigator information concerning the pattern of symptoms.

Some more detailed applications of principal components analysis are given in Section 4.8, but here we continue with their mathematical derivation.

4.2 Definition of Principal Components

The first principal component of the observations is that linear combination, y_1, of the original variables,

$$y_1 = a_{11}x_1 + a_{12}x_2 + \cdots + a_{1p}x_p \tag{4.1}$$

whose sample variance is greatest for all coefficients, $a_{11}, \ldots, a_{1p}$ (which we may write as the vector $\mathbf{a}_1$). Since the variance of y_1 could be increased without limit simply by increasing the elements of $\mathbf{a}_1$, a restriction must be placed on these coefficients; as we shall see later a sensible constraint is to require that the sum-of-squares of the coefficients, i.e. $\mathbf{a}_1'\mathbf{a}_1$, should be set at a value of unity.

But how useful is this artificial variate constructed from the observed variables? To answer this question one would first need to know the proportion of the total variance for which it accounted. If, for example, 80 per cent of the variation in an investigation involving six variables could be accounted for by a simple weighted average of the variable values, then almost all the variation could be expressed along a single continuum rather than in six-dimensional space. This would provide a highly parsimonious summary of the data that might be useful in later analyses.

The second principal component, y_2, is that linear combination

$$y_2 = a_{21}x_1 + a_{22}x_2 + \cdots + a_{2p}x_p \tag{4.2}$$

i.e.

$$y_2 = \mathbf{a}_2'\mathbf{x}$$

which has the greatest variance subject to the two conditions

$$\mathbf{a}_2'\mathbf{a}_2 = 1 \quad \text{(for the reason indicated previously)}$$

and

$$\mathbf{a}_2'\mathbf{a}_1 = 0 \quad \text{(so that } y_1 \text{ and } y_2 \text{ are uncorrelated)}$$

Similarly the jth principal component is that linear combination

$$y_j = \mathbf{a}_j'\mathbf{x} \tag{4.3}$$

which has greatest variance subject to

$$\mathbf{a}'_j\mathbf{a}_j = 1$$
$$\mathbf{a}'_j\mathbf{a}_i = 0 \qquad (i < j).$$

To find the coefficients defining the first principal component we need to choose the elements of $\mathbf{a}_1$ so as to maximise the variance of y_1 subject to the constraint, $\mathbf{a}'_1\mathbf{a}_1 = 1$. The variance of y_1 is given by

$$\text{Var}(y_1) = \text{Var}(\mathbf{a}'_1\mathbf{x})$$
$$= \mathbf{a}'_1\mathbf{S}\mathbf{a}_1 \tag{4.4}$$

where $\mathbf{S}$ is the variance–covariance matrix of the original variables. (This result is derived in Section 2.4.)

The standard procedure for maximising a function of several variables, subject to one or more constraints, is the method of *Lagrange multipliers*. Applying this technique to maximise the variance of y_1, as given by (4.4), subject to the constraint, $\mathbf{a}'_1\mathbf{a}_1 = 1$, leads to the solution that $\mathbf{a}_1$ is the eigenvector of $\mathbf{S}$ corresponding to the largest eigenvalue. (For details see Morrison, 1967 and Chatfield and Collins, 1980.) To determine the second component, the Lagrange multiplier technique is again used to maximise the variance of y_2, $\mathbf{a}'_2\mathbf{S}\mathbf{a}_2$, subject to the two constraints, $\mathbf{a}'_2\mathbf{a}_2 = 1$ and $\mathbf{a}'_2\mathbf{a}_1 = 0$. This leads to the solution that $\mathbf{a}_2$ is the eigenvector of $\mathbf{S}$ corresponding to its second largest eigenvalue. Similarly the jth principal component is defined by the eigenvector associated with the jth largest eigenvalue.

If the eigenvalues of $\mathbf{S}$ are $\lambda_1, \lambda_2, \ldots, \lambda_p$ then it is easy to show that by choosing $\mathbf{a}'_i\mathbf{a}_i = 1$ the variance of the ith principal component is given by λ_i. For example y_1 has variance given by (4.4); now since $\mathbf{a}_1$ is an eigenvector of $\mathbf{S}$ we know that

$$\mathbf{S}\mathbf{a}_1 = \lambda_1\mathbf{a}_1 \tag{4.5}$$

therefore (4.4) may be rewritten as

$$\text{Var}(y_1) = \mathbf{a}'_1\lambda_1\mathbf{a}_1 \tag{4.6}$$
$$= \lambda_1\mathbf{a}'_1\mathbf{a}_1$$
$$= \lambda_1 \quad (\text{since } \mathbf{a}'_1\mathbf{a}_1 = 1)$$

The total variance of the p principal components will equal the total variance of the original variables so that

$$\sum_{i=1}^{p} \lambda_i = \text{trace}(\mathbf{S}) \tag{4.7}$$

Consequently the jth principal component accounts for a proportion

$$t = \frac{\lambda_j}{\text{trace}(\mathbf{S})} \tag{4.8}$$

of the total variation in the original data and the first, say p_1, components $(p_1 < p)$ account for a proportion

$$T = \frac{\sum_{i=1}^{p_1} \lambda_i}{\text{trace}(\mathbf{S})} \tag{4.9}$$

of the total variation. Although the derivation of principal components given above has been in terms of the eigenvalues and eigenvectors of the covariance matrix, S, it is much more usual in practice to derive them from the corresponding quantities of the correlation matrix, R. The reasons are not difficult to appreciate if we imagine a set of multivariate data where the variables, $x_1, x_2, \ldots, x_p$, are of completely different types, for example lengths, temperatures, blood pressures, anxiety ratings etc. In such a case the structure of the principal components derived from the covariance matrix will depend upon the essentially arbitrary choice of units of measurement; additionally if there are large differences between the variances of $x_1, x_2, \ldots, x_p$, these variables whose variances are largest will tend to dominate the first few principal components. (An example illustrating this problem is given in Jolliffe, 1986.) Extracting the components as the eigenvectors of R which is equivalent to calculating the principal components from the original variables after each has been standardised to have unit variance overcomes this difficulty. It is important to realise, however, that the eigenvalues and eigenvectors of R will generally not be the same as those of S; indeed there is rarely any simple correspondence between the two and choosing to analyse R rather than S involves a definite but possibly arbitrary decision to make the variables 'equally important'.

4.3   Correlations of Variables and Components

The covariance of the observed variables with the jth principal component are easily found as follows:

$$\text{Cov}(\mathbf{x}, y_j) = \text{Cov}(\mathbf{x}, \mathbf{x}'\mathbf{a}_j) \tag{4.10}$$

$$= E(\mathbf{x}\mathbf{x}')\mathbf{a}_j \tag{4.11}$$

$$= \mathbf{S}\mathbf{a}_j \tag{4.12}$$

$$= \lambda_j \mathbf{a}_j \tag{4.13}$$

Consequently the covariance of variable i with component j is given by

$$\text{Cov}(x_i, y_j) = \lambda_j a_{ji} \tag{4.14}$$

and the correlation of variable i with component j is therefore

$$r_{x_i y_j} = \frac{\text{Cov}(x_i, y_j)}{\sigma_{x_i} \sigma_{y_j}} \tag{4.15}$$

$$= \frac{\lambda_j a_{ji}}{s_{ii}^{\frac{1}{2}} \sqrt{\lambda_j}} \tag{4.16}$$

$$= \frac{\sqrt{\lambda_j} a_{ji}}{s_{ii}^{\frac{1}{2}}} \tag{4.17}$$

If the components are extracted from the correlation matrix rather than the covariance matrix, then

$$r_{x_i y_j} = \sqrt{\lambda_j} a_{ji} \tag{4.18}$$

4.4   Rescaling Principal Components

If the vectors, $\mathbf{a}_1, \mathbf{a}_2, \ldots, \mathbf{a}_p$, which define the principal components are used to form a $p \times p$ matrix, $\mathbf{A} = [\mathbf{a}_1, \ldots, \mathbf{a}_p]$, and the eigenvalues, $\lambda_1, \ldots, \lambda_p$, arranged in a diagonal matrix, $\mathbf{\Lambda}$, then it is easy to show that the covariance matrix, S

of original variables is given by

$$S = A\Lambda A' \tag{4.19}$$

(We are assuming that the components have been extracted from **S** rather than **R**.)

By rescaling the vectors $\mathbf{a}_1, \ldots, \mathbf{a}_p$ so that the sum of squares of their elements is equal to the correspondence eigenvalue, λ_i, rather than unity, i.e. calculating $\mathbf{a}_i = \lambda_i^{\frac{1}{2}} \mathbf{a}_i$ then (4.19) may be written more simply as

$$S = (A^*)(A^*)' \tag{4.20}$$

where $A^* = [\mathbf{a}_1^*, \ldots, \mathbf{a}_p^*]$. The elements of A^* are such that the coefficients of the more important components are scaled to be generally larger than those of the less important components, which is intuitively reasonable. The rescaled vectors have a number of other advantages since the elements of A^* are analogous to *factor loadings* as we shall see in Chapter 13. In the case of components arising from a correlation matrix, the rescaled elements in A^* give, as we have shown in the previous section, correlations between the components and the original variables. Consequently when a principal components analysis is reported it is quite common to present the vectors $\mathbf{a}_1^*, \ldots, \mathbf{a}_p^*$ rather than $\mathbf{a}_1, \ldots, \mathbf{a}_p$.

4.5   Calculating Principal Component Scores

To obtain the scores for an individual on the derived principal components we could simply apply formulae (4.1), (4.2) and (4.3) to the individual original variable values. It is, however, generally more convenient to arrange matters so that the principal component scores have zero mean, by applying the vectors, $\mathbf{a}_1, \ldots, \mathbf{a}_p$ to the vector $(\mathbf{x}_i - \bar{\mathbf{x}})$, where $\mathbf{x}_i$ contains the original variable values for individual i and $\bar{\mathbf{x}}$ is the vector of mean values of the original variables. The component scores for individual i are thus given by

$$y_{i1} = \mathbf{a}_1'(\mathbf{x}_i - \bar{\mathbf{x}})$$
$$\vdots \tag{4.21}$$
$$y_{ip} = \mathbf{a}_p'(\mathbf{x}_i - \bar{\mathbf{x}})$$

How these principal component scores might be used will be described in later sections.

4.6   Choosing the Number of Components

As outlined above, principal components analysis is seen to be a technique for transforming a set of observed variables into a new set of variables which are uncorrelated with one another. The *total* variation in the original p variables is only accounted for by *all* p principal components. The usefulness of these components, however, stems from their property of accounting for the variance in decreasing proportions. So the important question arises as to how many components are needed to provide an adequate summary of a given data set?

Although a number of more formal techniques are available (see Jolliffe 1986), here we concentrate on informal, *ad hoc* rules of thumb which seek to answer this question and whose main justification is that they are intuitively plausible and that they often work in practice. The most common of these rules are as follows:

(1) Include just enough components to explain some relatively large percentage of the total variation. Figures between 70 and 90 per cent are usually suggested although this will generally become smaller as p or n increases.

(2) Exclude those principal components whose eigenvalues are less than the average, i.e. less than one if the components have been extracted from the correlation matrix. This rule was originally put forward by Kaiser (1958), but Jolliffe (1972) has suggested, on the basis of a number of simulation studies, that a more appropriate rule would be to exclude components whose associated eigenvalues are less than 0.7. (Again this assumes the use of the correlation rather than the covariance matrix.)

(3) Cattell (1965) suggests examination of a plot of λ_i against i and selecting as the number of components the value of i corresponding to an 'elbow' in the curve, this point being considered to be where 'large' eigenvalues cease and 'small' eigenvalues begin. Such a plot is generally known as a *scree* diagram. (Jolliffe, 1986, points out that an alternative to the scree graph, which is popular particularly in meteorology, is to plot $\log(\lambda_i)$ rather than λ_i against i; this is known as the *log-eigenvalue diagram*.)

Examples of the use of these criteria are given in Section 4.8.

4.7   A Simple Example of Principal Components Analysis

Suppose we have two variable x_1 and x_2 whose correlation matrix is given by

$$\mathbf{R} = \begin{pmatrix} 1.0 & r \\ r & 1.0 \end{pmatrix} \qquad (r > 0) \tag{4.22}$$

In order to find the principal components we first need to find the eigenvalues and eigenvectors of $\mathbf{R}$. The eigenvalues are roots of the equation

$$|\mathbf{R} - \lambda\mathbf{I}| = 0 \tag{4.23}$$

i.e.

$$(1 - \lambda)^2 - r^2 = 0 \tag{4.24}$$

giving eigenvalues $\lambda_1 = 1 + r$ and $\lambda_2 = 1 - r$. (Note that the sum of the eigenvalues is 2.) The eigenvector corresponding to λ_1 is obtained by solving

$$\mathbf{R}\mathbf{a}_1 = \lambda_1\mathbf{a}_1 \tag{4.25}$$
$$a_{11} + ra_{12} = (1 + r)a_{11} \tag{4.26}$$
$$ra_{11} + a_{12} = (1 + r)a_{12}$$

These two equations are identical and both reduce to the equation

$$a_{11} = a_{12} \tag{4.27}$$

If we now introduce the normalisation constraint, $\mathbf{a}_1'\mathbf{a}_1 = 1$, we find that

$$a_{11} = a_{12} = \frac{1}{\sqrt{2}} \tag{4.28}$$

Similarly we find the second eigenvector to be given by $a_{21} = \frac{1}{\sqrt{2}}$, $a_{22} = -\frac{1}{\sqrt{2}}$. Consequently the two principal components have the form

$$y_1 = \frac{1}{\sqrt{2}}(x_1 + x_2) \tag{4.29}$$

$$y_2 = \frac{1}{\sqrt{2}}(x_1 - x_2)$$

If $r < 0$ the order of the eigenvalues and hence of the principal components is reversed; if $r = 0$ the eigenvalues are both equal to 1 and any two components at right angles could be chosen. Two further points to note are

(1) There is an arbitrary sign in the choice of the elements of $\mathbf{a}_i$; it is customary to choose a_{i1} to be positive.
(2) The components do not depend on r, although the proportion of variance explained by each does change with r. As r tends to 1 the proportion of variance accounted for by y_1, namely $(1 + r)/2$, also tends to 1.

4.8 Numerical Examples

4.8.1 Drug Usage by American College Students

The majority of adult and adolescent Americans regularly use psychoactive substances during an increasing proportion of their life time. Various forms of licit and illicit psychoactive substance use are prevalent, suggesting that patterns of psychoactive substance taking are major parts of the individuals, behavioural repertory and have pervasive implications for the performance of other behaviours. In an investigation of these phenomena Huba, *et al* (1981) collected data on drug usage rates for 1634 students in the seventh through ninth grades in 11 schools in the greater metropolitan area of Los Angeles. Each participant completed a questionnaire about the number of times a particular substance had ever been used. The substances for which data were collected were as follows:

(1) Cigarettes
(2) Beer
(3) Wine
(4) Liquor
(5) Cocaine
(6) Tranquillisers
(7) Drug store medications used to get high
(8) Heroin and other opiates
(9) Marijuana
(10) Hashish
(11) Inhalents (glue, gasoline, etc.)
(12) Hallucinogenics (LSD, mescaline etc.)
(13) Amphetamine stimulants

Table 4.1   Correlation matrix for drug usage data

	Drug	1	2	3	4	5	6	7	8	9	10	11	12	13
1	Cigs	1												
2	Beer	447	1											
3	Wine	422	619	1										
4	Liquor	435	604	583	1									
5	Cocaine	114	068	053	115	1								
6	Tranq	203	146	139	258	349	1							
7	Drugst	091	103	110	122	209	221	1						
8	Heroin	082	063	066	097	321	355	201	1					
9	Mari	513	445	365	482	186	315	150	154	1				
10	Hash	304	318	240	368	303	377	163	219	534	1			
11	Inhal	245	203	183	255	272	323	310	288	301	302	1		
12	Hallu	101	088	074	139	279	367	232	320	204	368	340	1	
13	Amphet	245	199	184	293	278	545	232	314	394	467	392	511	1

(Decimal points omitted)

Table 4.2   Principal components for drug usage correlations in Table 4.1

	PC1	PC2	PC3	PC4	PC5	PC6	PC7	PC8	PC9	PC10	PC11	PC12	PC13
1	0·58	0·40	−0·06	0·01	0·27	0·38	0·10	−0·02	0·46	−0·08	−0·09	−0·13	−0·12
2	0·60	0·57	0·13	0·09	−0·15	−0·12	−0·09	0·04	−0·04	−0·06	−0·05	0·38	−0·30
3	0·55	0·56	0·21	0·13	−0·27	−0·13	−0·05	−0·06	0·07	−0·28	0·17	−0·09	0·34
4	0·66	0·46	0·05	0·06	−0·15	−0·13	0·00	−0·09	−0·15	0·40	−0·07	−0·30	−0·08
5	0·44	−0·41	0·05	0·53	0·38	−0·29	−0·24	−0·18	0·13	0·05	0·10	0·02	−0·00
6	0·61	−0·37	−0·17	0·08	−0·11	−0·04	0·44	−0·38	−0·06	−0·06	−0·28	0·11	0·08
7	0·37	−0·27	0·71	0·30	0·22	−0·22	0·27	0·16	0·05	0·00	−0·00	−0·02	−0·02
8	0·42	−0·45	−0·14	0·48	0·28	−0·32	0·13	0·39	−0·08	0·01	0·03	−0·04	−0·03
9	0·71	0·23	−0·23	−0·10	0·31	0·12	0·116	0·20	−0·10	0·23	0·13	0·24	0·28
10	0·69	−0·07	−0·35	−0·11	0·22	−0·20	−0·01	0·29	−0·26	−0·27	−0·16	−0·18	−0·07
11	0·58	−0·24	0·31	−0·18	0·07	0·42	−0·38	−0·28	−0·28	−0·05	−0·04	0·01	0·02
12	0·52	−0·47	−0·11	−0·26	−0·31	−0·12	−0·32	0·15	0·36	0·13	−0·17	0·07	0·10
13	0·69	−0·33	−0·23	−0·24	−0·18	−0·02	0·11	−0·14	0·04	−0·05	0·45	−0·06	−0·18
EV	4·38	2·04	0·95	0·82	0·76	0·69	0·64	0·61	0·56	0·40	0·40	0·39	0·37

Responses were recorded on a five-point scale 1. Never tried, 2. Only once, 3. A few times, 4. Many times, 5. Regularly. The correlations between the usage rates of the 13 substances are shown in Table 4.1. In the original paper these correlations were used as the basis for testing the adequacy of various models of drug taking, a point to which we shall return in Chapters 13 and 14. Here we use the derived correlation matrix to illustrate the use of principal components analysis.

The coefficients defining the 13 principal components of these data are given in Table 4.2. These coefficients are scaled as described in Section 4.4, so that they represent correlations between observed variables and derived components. The eigenvalues corresponding to each coefficient are also shown in Table 4.2. A scree diagram of these components is given in Figure 4.1. This clearly indicates that there are two major components in the data, a fact which is further emphasised by examining the eigenvalues, the first two of which have values greater than one.

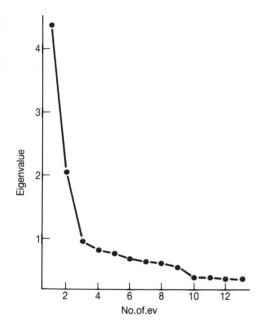

Figure 4.1   Scree diagram for drug usage data

   Thc analysis suggests that the first two principal components scores for each individual might act as an adequate summary of the original 13 scores in any further analyses of the data; these two components account for nearly 50 % of the total variation of the original variables.

   Such a summary would clearly be more useful if the two components could be given some meaningful interpretation. Here this is perhaps not too difficult. The first component appears to measure overall drug usage as might be expected since the correlations between all 13 variables are positive. The second components contrasts 'legal' with 'illegal' substances (with the exception of marijuana which has the same sign coefficient as the legal substances); alternatively this component could be seen as contrasting 'soft' and 'hard' drug usage. So we see that after overall usage has been accounted for, the main source of variation is between individuals who consume the two different types of substance.

   Most users of principal components analysis search for an interpretation of the components in this fashion, but this is not without its difficulties and critics, and the following quotation from Marriott (1974) should act as a salutory warning against the dangers of over interpretation.

   It must be emphasised that no mathematical method is, or could be, designed to give physically meaningful results. If a mathematical expression of this sort has an obvious physical meaning, it must be attributed to a lucky chance, or to the fact that the data have a strongly marked structure that shows up in the analysis. Even in the latter case, quite small sampling fluctuations can upset the interpretation;

for example the first two principal components may appear in reverse order, or may become confused altogether. Reification, then, requires considerable skill and experience if it is to give a true picture of the physical meaning of the data.

4.8.2   Using Principal Component Scores to Display Multivariate Data

Attempting to identify components in the way described above for the drug usage data can be (and often is) a very subjective procedure. Indeed the identification of components is to a considerable extent arbitrary in that completely different results may be produced by analysing the covariance matrix rather than the correlation matrix. This suggests that it could be misleading to try to attach too much meaning to components in many situations, and perhaps a more important objective which this type of analysis has is the possibility for reducing the dimensionality of the data as a prelude to further investigations, when the question of trying to interpret the components *en route* becomes less important. One area where this may be very useful is that of multiple regression, which we shall discuss in Chapter 8. Another often very helpful application of principal components analysis, not involving any need for interpretation of the components, is in providing low-dimensional plots of the data, which can be an aid in identifying outlying observations, clusters of similar observations, and so on. To illustrate this use of the technique we shall use the athletics data shown in Table 4.3, which gives the record times for various track events in 55 different countries. The correlations between the times are shown in Table 4.4. These show a not unexpected pattern with higher correlations between events of similar length. The coefficients defining the first two principal components are shown in Table 4.5. The first component clearly represents a measure of the overall athletic prowess of a country; countries with above average times on the majority of events tend to have high scores on this component. The second component largely contrasts 'sprints' with middle and long distance times. Countries particularly poor at the shorter distances tend to have high positive scores on this component, and those with poor middle-long distant records have high negative scores.

A plot of the 55 countries in the space of the first two principal components is given in Figure 4.2. Some of the points to note about this diagram are

(1)   Both the Cook Islands and Western Samoa have high scores on the first component reflecting the relatively low overall athletic standard in these two countries for the events considered. The two countries are however differentiated by their scores on the second component; here the Cook Islands have a high positive score and Western Samoa a high negative score. This reflects the particularly low sprinting standard in the former country and the relatively poor middle–long distance performance in the latter.

(2)   Bermuda, Malaysia, Singapore and Thailand have very similar record profiles and a higher standard of sprinting than middle distance running.

(3)   The second component separates out countries of similar overall athletic prowess into those where sprinting predominates such as the USA, Italy and Brazil, and those with higher middle distance standards such as Portugal and Norway.

For interest and comparison a 'faces' representation of these data is given in Figure 4.3.

We shall return to these data in Chapter 6.

A further example appears in Figure 4.4, which shows a set of five-dimensional observations plotted in the space of its first two principal components. The presence of an outlying observation is clearly indicated, and this observation would need to be examined to decide whether or not it should be removed before attempting further analyses.

Some other interesting examples of the use of principal component plots are given in Gower and Digby (1981). A number of other techniques that are related to principal components analysis and may also be used to obtain informative displays of multivariate data are discussed by Gabriel (1981).

4.8.3 Selecting a Subset of Variables

Although the first few principal component scores for each individual may provide a very useful summary of a set of multivariate data, *all* the original variables are needed in their computation. In many cases an investigator might be happier with determining a subset of the original variables which contain virtually all the information available in the complete set of these variables. Jolliffe (1970, 1972, 1973) in a series of papers discusses a number of methods for selecting subsets of variables, several of which are based on principal components analysis.

One of these is to use one or other of the criteria discussed in Section 4.6 to decide on the number of components which account for most of the variation in the original variables; this number is taken to indicate the effective dimensionality of these variables. Suppose p^* components are so chosen. One variable is then associated with each of the components, namely that variable not already chosen which has the highest coefficient in absolute value on the component. These p^* variables are the required subset. To illustrate this procedure we shall return again to the drug usage data discussed in Section 4.8.1. If now we use the eigenvalue criteria suggested by Jolliffe (i.e. retain components with eigenvalues greater than 0.7), five components are kept. Examining the coefficients defining these components (see Table 4.2), we find the subset of 5 variables chosen according to the rule described above is

(9) Marijuana
(2) Beer
(7) Drugstore
(5) Cocaine
(12) Hallucinogenic

4.9 Geometrical Interpretation of Principal Components Analysis

In geometrical terms it is easy to show that the first principal component defines the line of best fit (in the least squares sense) to the p-dimensional observations in the sample. These observations may therefore be represented in one dimension by taking their projection onto this line, i.e. finding their first principal component score. If the points happened to be collinear this

Table 4.3   Athletic records for 55 countries

Country	1	2	3	4	5	6	7	8
Argentina	10·39	20·81	46·84	1·81	3·70	14·04	29·36	137·72
Australia	10·31	20·06	44·84	1·74	3·57	13·28	27·66	128·30
Austria	10·44	20·81	46·82	1·79	3·60	13·26	27·72	135·90
Belgium	10·34	20·68	45·04	1·73	3·60	13·22	27·45	129·95
Bermuda	10·28	20·58	45·91	1·80	3·75	14·68	30·55	146·62
Brazil	10·22	20·43	45·21	1·73	3·66	13·62	28·62	133·13
Burma	10·64	21·52	48·30	1·80	3·85	14·45	30·28	139·95
Canada	10·17	20·22	45·68	1·76	3·63	13·55	28·09	130·15
Chile	10·34	20·8	46·20	1·79	3·71	13·61	29·30	134·03
China	10·51	21·04	47·30	1·81	3·73	13·90	29·13	133·53
Colombia	10·43	21·05	46·10	1·82	3·74	13·49	27·88	131·35
Cook Is	12·18	23·2	52·94	2·02	4·24	16·70	35·38	164·70
Costa Rica	10·94	21·9	48·66	1·87	3·84	14·03	28·81	136·58
Czech	10·35	20·65	45·64	1·76	3·58	13·42	28·19	134·32
Denmark	10·56	20·52	45·89	1·78	3·61	13·50	28·11	130·78
Dom Rep	10·14	20·65	46·80	1·82	3·82	14·91	31·45	154·12
Finland	10·43	20·69	45·49	1·74	3·61	13·27	27·52	130·87
France	10·11	20·38	45·28	1·73	3·57	13·34	27·97	132·30
GDR	10·12	20·33	44·87	1·73	3·56	13·17	27·42	129·92
FRG	10·16	20·37	44·50	1·73	3·53	13·21	27·61	132·23
GB	10·11	20·21	44·93	1·70	3·51	13·01	27·51	129·13
Greece	10·22	20·71	46·56	1·78	3·64	14·59	28·45	134·60
Guatemala	10·98	21·82	48·40	1·89	3·80	14·16	30·11	139·33
Hungary	10·26	20·62	46·02	1·77	3·62	13·49	28·44	132·58
India	10·60	21·42	45·73	1·76	3·73	13·77	28·81	131·98
Indonesia	10·59	21·49	47·80	1·84	3·92	14·73	30·79	148·83
Ireland	10·61	20·96	46·30	1·79	3·56	13·32	27·81	132·35
Israel	10·71	21·00	47·80	1·77	3·72	13·66	28·93	137·55
Italy	10·01	19·72	45·26	1·73	3·60	13·23	27·52	131·08
Japan	10·34	20·81	45·86	1·79	3·64	13·41	27·72	128·63
Kenya	10·46	20·66	44·92	1·73	3·55	13·10	27·80	129·75
Korea	10·34	20·89	46·90	1·79	3·77	13·96	29·23	136·25
P Korea	10·91	21·94	47·30	1·85	3·77	14·13	29·67	130·87
Luxemburg	10·35	20·77	47·40	1·82	3·67	13·64	29·08	141·27
Malaysia	10·40	20·92	46·30	1·82	3·80	14·64	31·01	154·10
Mauritus	11·19	22·45	47·70	1·88	3·83	15·06	31·77	152·23
Mexico	10·42	21·30	46·10	1·80	3·65	13·46	27·95	129·20
Netherlands	10·52	29·95	45·10	1·74	3·62	13·36	27·61	129·02
NZ	10·51	20·88	46·10	1·74	3·54	13·21	27·70	128·98
Norway	10·55	21·16	46·71	1·76	3·62	13·34	27·69	131·48
Png	10·96	21·78	47·90	1·90	4·01	14·72	31·36	148·22
Philippines	10·78	21·64	46·24	1·81	3·83	14·74	30·64	145·27
Poland	10·16	20·24	45·36	1·76	3·60	13·29	27·89	131·58
Portugal	10·53	21·17	46·70	1·79	3·62	13·13	27·38	128·65
Rumania	10·41	20·98	45·87	1·76	3·64	13·25	27·67	132·50
Singapore	10·38	21·28	47·40	1·88	3·89	15·11	31·32	157·77
Spain	10·42	20·77	45·98	1·76	3·55	13·31	27·73	131·57
Sweden	10·25	20·61	45·63	1·77	3·61	13·29	27·94	130·63
Switzerland	10·37	20·45	45·78	1·78	3·55	13·22	27·91	131·20
Tapei	10·59	21·29	46·80	1·79	3·77	14·07	30·07	139·27
Thailand	10·39	21·09	47·91	1·83	3·84	15·23	32·56	149·90
Turkey	10·71	21·43	47.60	1·79	3·67	13·56	28·58	131·50
USA	9·93	19·75	43.86	1·73	3·53	13·20	27·43	128·22
USSR	10·07	20·00	44.60	1·75	3·59	13·20	27·53	130·55
W Samoa	10·82	21·86	49.00	2·02	4·24	16·28	34·71	161·83

Event: (1) 100 m (s), (2) 200 m (s), (3) 400 m (s), (4) 800 m (min), (5) 1500 m (min), (6) 5000 m (min), (7) 10000 m, (8) Marathon (min).

Table 4.4   Correlations for athletics data

	100 m	200 m	400 m	800 m	1500 m	5000 m	10000 m	Marathon
100 m	1·00							
200 m	0·91	1·00						
400 m	0·83	0·84	1·00					
800 m	0·72	0·76	0·81	1·00				
1500 m	0·68	0·75	0·82	0·78	1·00			
5000 m	0·62	0·69	0·78	0·74	0·91	1·00		
10000 m	0·63	0·68	0·78	0·74	0·93	0·97	1·00	
Marathon	0·52	0·60	0·70	0·67	0·86	0·93	0·94	1·00

Table 4.5   First two principal components for athletics data

	PC1	PC2
100 m	0·82	0·50
200 m	0·86	0·41
400 m	0·92	0·21
800 m	0·87	0·15
1500 m	0·94	−0·16
5000 m	0·93	−0·30
10000 m	0·94	−0·31
Marathon	0·87	−0·42
Eigenvalue	6·41	0·89

representation would account completely for the variation in the data, and the sample covariance matrix would have only non-zero eigenvalue. In practice, of course, the points are unlikely to be collinear, and an improved representation would be given by projecting them onto the plane of best fit, this being defined by the first two principal components. Similarly the first p^* components give the best fit in p^* dimensions. If the observations fit exactly into a space of p^* dimensions, this will be indicated by the presence of $p - p^*$ zero eigenvalues of the covariance matrix. This would imply the presence of $p - p^*$ linear constraints on the variables. These constraints are sometimes referred to as *structural relationships*.

4.10   Correspondence Analysis

Correspondence analysis attempts to display graphically the relationships in a two-way table of counts (such as that seen in Table 4.6), by deriving *coordinates* representing the row categories and column categories of the table.

The correspondence analysis coordinates are analogous to those derived from principal components of continuous data, except that they partition the total χ^2 value used in testing independence, rather than the total variance. In this section only a brief account of the method is given ; a full account of correspondence analysis is available in Greenacre (1983).

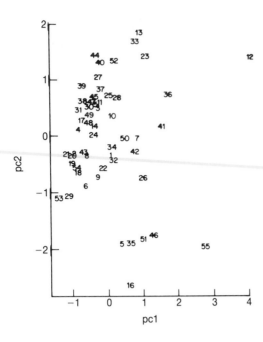

Figure 4.2 Principal components plot for athletic records data

The well-known chi-square test for independence in a two-dimensional *contingency table* such as Table 4.6 consists of first finding estimated expected values for each cell of the table under the hypothesis of independence and then comparing these with corresponding observed values using the test statistic, X^2 given by

$$X^2 = \sum_{i=1}^{I} \sum_{j=1}^{J} \frac{(n_{ij} - E_{ij})^2}{E_{ij}} \qquad (4.30)$$

where n_{ij}, $i = 1,\dots,I$, $j = 1,\dots,J$ are the observed values and E_{ij} the corresponding expected values: I is the number of rows, and J the number of columns of the table. The E_{ij} are calculated from the appropriate marginal totals as

$$E_{ij} = \frac{n_{i.}n_{.j}}{n_{..}} \qquad (4.31)$$

Under the hypothesis of independence, X^2 has a chi-squared distribution with $(I-1)(J-1)$ degrees of freedom. In the case of the eye colour, hair colour data in Table 4.6, the test statistic takes the value 1240 with 12 d.f., indicating a highly significant departure from independence.

It is in the examination of the specific reasons for departure from independence that correspondence analysis becomes useful. The method essentially consists of finding the singular value decomposition of the matrix, **C**, containing the individual components of the chi-square statistic in (4.30), i.e.

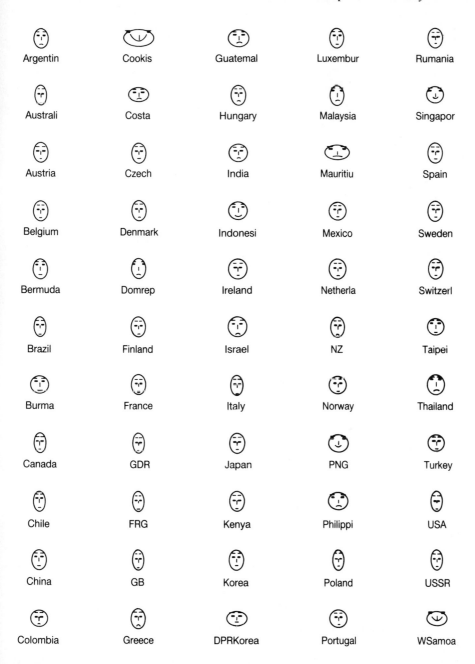

Figure 4.3   Faces representation of athletic records data

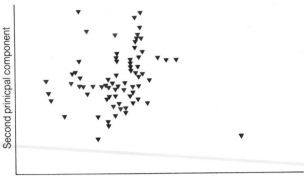

Figure 4.4 Plot of a set of five-dimensional observations in the space of the first two principal components

Table 4.6 Eye colour, hair colour data

Eye Colour	Hair colour				
	Fair	Red	Medium	Dark	Black
Light	688	116	584	188	4
Blue	326	38	241	110	3
Medium	343	84	909	412	26
Dark	98	48	403	681	81

$c_{ij} = \frac{(n_{ij} - E_{ij})^{\frac{1}{2}}}{\sqrt{E_{ij}}}$. Such a decomposition involves finding matrices $\mathbf{U}$, $\mathbf{V}$ and $\mathbf{\Lambda}$ (diagonal) such that

$$\mathbf{C} = \mathbf{U}\mathbf{\Lambda}\mathbf{V}' \tag{4.32}$$

where $\mathbf{U}$ contains the eigenvectors of $\mathbf{CC}'$ and $\mathbf{V}$ the eigenvectors of $\mathbf{C}'\mathbf{C}$. From (4.32) we find that the elements of $\mathbf{C}$ (i.e. the *residuals* from the independence model), can be written as

$$c_{ij} = \sum_{k=1}^{R} \delta_k^{\frac{1}{2}} u_{ik} v_{jk}, \quad i = 1,\ldots,I, \quad j = 1,\ldots,J \tag{4.33}$$

where $R = \min(I - 1, J - 1)$ is the rank of the matrix $\mathbf{C}$, and u_{ik} and v_{jk} are the elements of the kth column of $\mathbf{U}$ and the kth column of $\mathbf{V}$ respectively; $\delta_1,\ldots,\delta_R$ are the eigenvalues of $\mathbf{CC}'$, so that

$$\text{trace}(\mathbf{CC}') = \sum_{k=1}^{R} \delta_k = \sum_{i=1}^{I} \sum_{j=1}^{J} c_{ij}^2 = X^2 \tag{4.34}$$

Most applications of correspondence analysis use the first two columns of $\mathbf{U}$, $\mathbf{u}_1$ and $\mathbf{u}_2$ and the first two columns of $\mathbf{V}$, $\mathbf{v}_1$ and $\mathbf{v}_2$ to provide a graphical display of the c_{ij}, with the entries in $\mathbf{u}_1$ and $\mathbf{u}_2$ providing two-dimensional

Table 4.7 Correspondence analysis results for eye colour, hair colour data

Eye colour	u_1	u_2	Hair colour	v_1	v_2
LF	−0·535	−0·276	FH	−0·633	−0·521
BE	−0·327	−0·348	RH	−0·120	−0·064
ME	0·043	0·810	MH	−0·059	0·756
DE	0·778	−0·381	DH	0·670	−0·304
			BH	0·362	−0·245

δ_1	δ_2	δ_3	δ_4
1073·3	162·12	4·6	0·0

coordinates for points representing the row categories of the table, and those in $\mathbf{v}_1$ and $\mathbf{v}_2$ providing corresponding coordinates for the column categories. How well the two-dimensional coordinates represent the residuals can be judged by the size of the first two eigenvalues of $\mathbf{CC'}$ relative to the remainder.

An explanation of how to interpret the derived coordinates is made simpler if we consider the situation where the first eigenvalue is dominant so that the residuals from the independence model are well represented by the one-dimensional coordinates given by $\mathbf{u}_1$ and $\mathbf{v}_1$. In such a case we have that

$$c_{ij} \approx \delta_1^{\frac{1}{2}} u_{i1} v_{j1} \qquad (4.35)$$

From (4.35) we see that when u_{i1} and v_{j1} are both large and positive (or both large and negative) then c_{ij} will be large and positive indicating a positive association between row i and column j of the table. Similarly when u_{i1} and v_{j1} are large but have different signs, the ith row and jth column have a negative association. Finally when the product $u_{i1}v_{j1}$ is near zero, the association between the ith row and jth column is low. An important point which should be stressed is that it is *not* the closeness of a row point to a column point that determines their degree of association but the *comparison* of their distances from the origin.

To illustrate how the method works in practice we shall use the eye colour, hair colour data in Table 4.6. The results from applying correspondence analysis are shown in Table 4.7. Plotting the results gives Figure 4.5. It is this graphical display which is the most attractive feature of correspondence analysis. It allows a direct visualization of how eye colours are associated with hair colours. For example, it is clear from Figure 4.5 that there is a large positive association (large, positive c_{ij}) between dark eyes and black or dark hair, and a large negative association (large, negative c_{ij}), between dark eyes and fair hair.

A larger and perhaps more interesting example is provided by the data in Table 4.8 taken with permission from Hartigan (1975) . These data give the percentage of all households with various foods in the home at the time of survey.

Here the first two eigenvalues account for only 58.6% of the chi-square value, and so the corresponding two-dimensional representation will not tell the whole story about the structure in the table. Nevertheless the plot (given in Figure 4.6) does give an intuitively appealing picture (although this perhaps

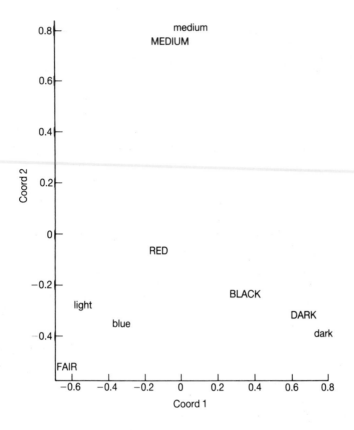

Figure 4.5   Correspondence analysis solution for eye colour (lower case)/hair colour (capitals) data

depends on one's *a priori* prejudices about the type of food eaten in particular countries). For example, Scandinavian countries (SW, DK, NY and FD) are associated with foods such as crispbread and frozen fish, Spain and Italy with olive oil and the UK and Ireland with tinned soup and tinned fruit. A more detailed interpretation is left as an exercise for the reader.

4.11   Summary

Principal components analysis looks for a few linear combinations of the original variables that can be used to summarise a data set, losing in the process as little information as possible. The derived variables may be used in a variety of ways, in particular for simplifying later analyses and providing informative plots of the data. The method consists essentially of transforming a set of correlated variables to a new set of variables which are uncorrelated. Consequently it should be noted that if the original variables are themselves almost uncorrelated there is little point in carrying out the principal compo-

Table 4.8 Percentages of households with various types of food
(Taken with permission from Hartigan, 1975)

	WG	IT	FR	NS	BM	LG	GB	PL	AA	SD	SW	DK	NY	FD	SP	ID
GC	90	82	88	96	94	97	27	72	55	73	97	96	92	98	70	13
IC	49	10	42	62	38	61	86	26	31	72	13	17	17	12	40	52
TB	88	60	63	98	48	86	99	77	61	85	93	92	83	84	40	99
SS	19	2	4	32	11	28	22	2	15	25	31	35	13	20	0	11
BP	57	55	76	62	74	79	91	22	29	31	0	66	62	64	62	80
SP	51	41	53	67	37	73	55	34	33	69	43	32	51	27	43	75
ST	19	3	11	43	25	12	76	1	1	10	43	17	4	10	2	18
IP	21	2	23	7	9	7	17	5	5	17	39	11	17	8	14	2
FF	27	4	11	14	13	26	20	20	15	19	54	51	30	18	23	5
VF	21	2	5	14	12	23	24	3	11	15	45	42	15	12	7	3
AF	81	67	87	83	76	85	76	22	49	79	56	81	61	50	59	57
OF	75	71	84	89	76	94	68	51	42	70	78	72	72	57	77	52
FT	44	9	40	61	42	83	89	8	14	46	53	50	34	22	30	46
JS	71	46	45	81	57	20	91	16	41	61	75	64	51	37	38	89
CG	22	80	88	15	29	91	11	89	51	64	9	11	11	15	86	5
BR	91	66	94	31	84	94	95	65	51	82	68	92	63	96	44	97
ME	85	24	47	97	80	94	94	78	72	48	32	91	94	94	51	25
OO	74	94	36	13	83	84	57	92	28	61	48	30	28	17	91	31
YT	30	5	57	53	20	31	11	6	13	48	2	11	2	0	16	3
CD	26	18	3	15	5	24	28	9	11	30	93	34	62	64	13	9

GC = ground coffee, IC = instant coffee, TB = tea or tea bags, SS = sugarless sweet, BP = packaged biscuits, SP = soup (packages), ST = soup (tinned), IP = instant potatoes, FF = frozen fish, VF = frozen vegetables, AF = fresh apples, OF = fresh oranges, FT = tinned fruit, JS = jam (shop), CG = garlic clove, BR = butter, ME = margarine, OO = olive, corn oil, YT = yogurt, CD = crispbread.

nents analysis, since it will simply find components which are close to original variables but arranged in decreasing order of variance. Finally it is important to restress that principal components is, essentially, a straightforward mathematical technique for an orthogonal rotation to principal axes (see Chapter 2). If this is remembered, it might make less likely the often exaggerated claims made for the results obtained from such an analysis.

Correspondence analysis is analogous to principal components but operates on categorical data arranged in the form of the contingency tables. Most often it is applied to two-dimensional tables to obtain a graphical representation of the residuals from the independence model. There are however related techniques, — for example, *multiple correspondence analysis* — which can be applied to more complex tables. For details see Greenacre (1983).

Exercises

4.1 Suppose that $\mathbf{x}' = [x_1, x_2]$ is such that $x_2 = 1 - x_1$ and $x_1 = 1$ with probability p, and $x_1 = 0$ with probability $q = 1 - p$. Find the covariance matrix of $\mathbf{x}$ and its eigenvalues and eigenvectors.

4.2 MacDonnell (1902) obtained measurements of seven physical characteristics for each of 3000 criminals. The corresponding correlation matrix is shown in Table 4.9.

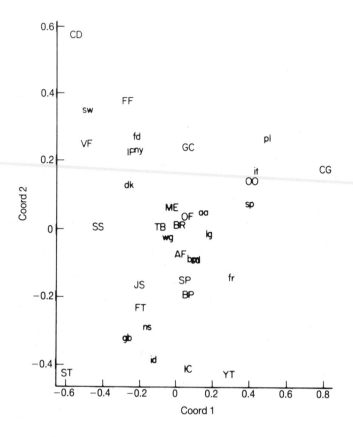

Figure 4.6 Correspondence analysis solution for European food data

The seven variables measured were (1) Head length, (2) head breadth, (3) face breadth, (4) left finger length, (5) left forearm, (6) left foot, (7) height. Find the principal components of these data and interpret the results.

4.3   The eigenvectors of a covariance matrix, **S**, scaled so that their sums of squares are equal to the corresponding eigenvalue are $c_1, \ldots, c_p$ show that

$$\mathbf{S} = \mathbf{c}_1 \mathbf{c}_1' + \cdots + \mathbf{c}_p \mathbf{c}_p'$$

(This result is known as the spectral decomposition of **S**.)

4.4   The data given in Table 4.10 show the distribution of episodes of particular psychiatric disorders in patients registered with 15 general practices. Compare the results of a principal components analysis on the raw data with those obtained after standardising in some way for practice size.

Table 4.9

Variables	1	2	3	4	5	6	7
1	1·000						
2	0·402	1·000					
3	0·395	0·618	1·000				
4	0·301	0·150	0·321	1·000			
5	0·305	0·135	0·289	0·846	1·000		
6	0·339	0·206	0·363	0·759	0·797	1·000	
7	0·340	0·183	0·345	0·661	0·800	0·736	1·000

Table 4.10

Practice	Size	R1	R2	R3
1	2519	221	68	6
2	1504	281	184	0
3	2161	234	206	3
4	4176	478	608	9
5	1480	76	251	3
6	2125	386	142	4
7	6514	1122	742	21
8	1820	208	398	9
9	2671	409	282	17
10	4220	314	429	11
11	2377	72	148	2
12	5009	566	305	7
13	2037	241	207	0
14	1759	290	277	2
15	1767	248	178	3

4.5 Find the principal components of the following correlation matrix, and compare how the 1 and 2 component solutions reproduce the matrix.

$$\mathbf{R} = \begin{pmatrix} 1 & & \\ 0.6579 & 1 & \\ 0.0034 & -0.0738 & 1 \end{pmatrix}$$

4.6 The data given in Table 4.11 concern counts of suicides cross-classified by method, sex, and age. Carry out a separate correspondance analysis for males and females and compare your results.

Table 4.11 Cause of death

Age Group	M_1	M_2	M_3	M_4	M_5	M_6
Men						
10–40	3983	1218	4555	1550	550	1248
40–70	3996	826	7971	1689	517	828
70+	938	45	3160	334	268	147
Women						
10–40	2593	153	956	141	407	383
40–70	4507	136	4528	264	715	601
70	1548	29	1856	77	383	106

Method
M_1 = suicide by solid or liquid matter; M_2 = suicide by gas, M_3 = suicide by hanging, strangling, suffocating or drowning, M_4 = suicide by guns, knives and explosives, M_5 = suicide by jumping, M_6 = suicide by other methods.

5

Multidimensional Scaling

5.1 Introduction

Apart from the raw multivariate data matrix, **X**, another frequently encountered type of data in the behavioural sciences is the *proximity matrix*, arising
either directly from experiments in which subjects are asked to assess the
similarity of two stimuli, or indirectly, as a measure of the correlation or
covariance of the two stimuli derived from their raw profile data. The investigator collecting such data is interested primarily in uncovering whatever
structure or pattern may be present in the observed proximity matrix, and the
subject of this chapter is one particularly powerful class of techniques which
may prove extremely useful in this search. Members of the class are generally
known as *multidimensional scaling techniques*, and the underlying purpose that
they share, despite their apparent diversity, is to represent the structure in the
proximity matrix by a simple geometrical model or picture.

A geometrical or spatial model for the observed proximity matrix consists
of a set of points, $x_1, x_2, \ldots, x_n$ in d dimensions (each point representing one
of the items or stimuli under investigation) and a measure of distance between
pairs of points. The object of multidimensional scaling is to determine both
the dimensionality of the model (that is the value of d) and the position of
the points in the resulting d-dimensional space, so that there is, in some sense,
maximum correspondence between the observed proximities and the interpoint
distances. In general terms this simply means that the larger the dissimilarity
between two stimuli (or the smaller their similarity), the further apart should
be the points representing the stimuli in the geometrical model. However, in
practice more explicit measures of how the proximities agree with the distances
are needed, and a variety of such measures have been suggested, giving rise to
a variety of multidimensional scaling techniques (see Section 5.4).

A number of interpoint distance measures are possible but by far the most
commonly used is *Euclidean distance*. This measure is illustrated for two
dimensions in Figure 5.1, and for d dimensions is given by

$$d_{ij} = \left[\sum_{k=1}^{d} (x_{ik} - x_{jk})^2 \right]^{\frac{1}{2}} \tag{5.1}$$

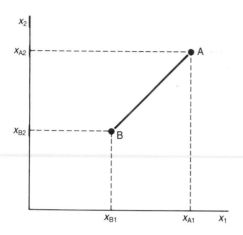

Figure 5.1   Euclidean distance in two dimensions

where $x_{i1}, \ldots, x_{id}$ and $x_{j1}, \ldots, x_{jd}$ are elements of vectors $\mathbf{x}_i$ and $\mathbf{x}_j$ respectively. Some other possible distance measures will be considered briefly in Section 5.7.

5.2   Simple Examples of Multidimensional Scaling

To clarify what the techniques of multidimensional scaling attempt to achieve, and before describing any technique in detail, we shall in this section look at a few simple examples, beginning with the data shown in Table 5.1. This is part of a data set collected by Rothkopf (1957); each entry in the table gives the number of times, expressed as a percentage, that pairs of Morse code signals for two numbers were declared to be the same by 598 subjects. Application of classical multidimensional scaling (see Section 5.3) results in the two-dimensional diagram shown in Figure 5.2. At the simplest level the purpose of this plot is to observe which signals were 'alike' (they are represented by points which are close together), and which were 'not alike' (they are represented by points far from each other). The plot should also indicate the general interrelationships between signals. A possible interpretation of the two axes is that the horizontal one is related to the increasing number of dots in a signal, whereas the vertical one is related to the ordering of dots and dashes in the signal. We shall return to this example later.

As a further example consider the data shown in Table 5.2, which results from averaging the ratings of eighteen students on the degree of overall similarity between twelve nations on a scale ranging from 1 indicating 'very different' to 9 for 'very similar.' No instructions were given concerning the characteristics on which these similarity judgements were to be made. (Further details of the study are given in Kruskal and Wish, 1978.)

The two-dimensional solution given by non-metric multidimensional scaling (see Section 5.4) appears in Figure 5.3. Kruskal and Wish's subjective interpretation of this solution, obtained by a rotation of the axes (see next section), is also shown.

Table 5.1   Number of times (%) that pairs of Morse code signals for two numbers were declared to be the same by 598 subjects (part of data in Rothkopf, 1957)

	1	2	3	4	5	6	7	8	9	0
(·----) 1	84									
(··---) 2	62	89								
(···--) 3	16	59	86							
(····-) 4	6	23	38	89						
(·····) 5	12	8	27	56	90					
(-····) 6	12	14	33	34	30	86				
(--···) 7	20	25	17	24	18	65	85			
(---··) 8	37	25	16	13	10	22	65	88		
(----·) 9	57	28	9	7	5	8	31	58	91	
(-----) 0	52	18	9	7	5	18	15	39	79	94

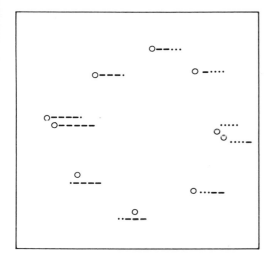

Figure 5.2   Two-dimensional solution from classical multidimensional scaling applied to the data in Table 5.1. (Adapted from Mardia *et al.*, 1979, courtesy of Academic Press)

5.3   Classical Multidimensional Scaling

Given the coordinates of n points in p-dimensional Euclidean space it is a simple matter to calculate the Euclidean distance d_{ij} between each pair of points. This can be done directly from the data matrix $\mathbf{X}$ using equation (5.1), or via the $(n \times n)$ matrix $\mathbf{B} = \mathbf{XX}'$ since

$$b_{ij} = \sum_{k=1}^{d} x_{ik} x_{jk} \tag{5.2}$$

so that

$$d_{ij}^2 = b_{ii} + b_{jj} - 2b_{ij} \tag{5.3}$$

Table 5.2 Matrix of mean similarity ratings for eighteen students for twelve nations (Adapted from Kruskal and Wish, 1978, courtesy of Sage Publications)

	BRZ	ZAI	CUB	EGY	FRA	IND	ISR	JPN	CHI	USSR	USA	YUG
Brazil	—											
Zaire	4·83	—										
Cuba	5·28	4·26	—									
Egypt	3·44	5·00	5·17	—								
France	4·72	4·00	4·11	4·78	—							
India	4·50	4·83	4·00	5·83	3·44	—						
Israel	3·83	3·33	3·61	4·67	4·00	4·11	—					
Japan	3·50	3·39	2·94	3·84	4·11	4·50	4·83	—				
China	2·39	4·00	5·50	4·39	3·67	4·11	3·00	4·17	—			
USSR	3·06	3·39	5·44	4·39	5·06	4·50	4·17	4·61	5·72	—		
USA	5·39	2·39	3·17	3·33	5·94	4·28	5·94	6·06	2·56	5·00	—	
Yugslav	3.17	3·50	5·11	4·28	4·72	4·00	4·44	4·28	5·06	6·67	3·56	—

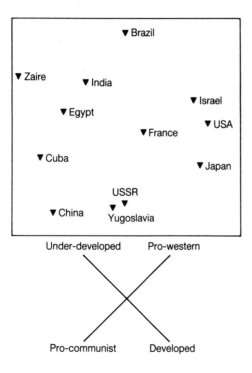

Figure 5.3 Two-dimensional solution from non-metric multidimensional scaling applied to the data in Table 5.2. (Adapted from Kruskal and Wish, 1978, courtesy of Sage Publications)

Now consider the inverse problem and suppose that we know the distances and wish to determine the coordinates. First we must note that there is no unique representation which will give rise to the distances, since these are unchanged by shifting the whole configuration of points from one place to another, or by rotation or reflection of the configuration. In other words, we cannot determine either the location or the orientation of the configuration. The location problem is usually overcome by placing the mean vector of the configuration at the origin. The orientation problem means that any configuration we obtain may be subjected to an arbitrary orthogonal transformation, i.e. a rigid rotation plus possibly a reflection. Such transformations are often used to facilitate the interpretation of solutions, as we have already seen in the previous section and will meet again in later parts of this chapter.

The procedure used to find the required coordinates consists of two stages, the first of which involves finding the elements of the matrix $\mathbf{B}$ introduced above, in terms of the known Euclidean distances, and the second of which consists of simply factorising $\mathbf{B}$ in the form $\mathbf{XX'}$. To get b_{ij} in terms of d_{ij} involves inverting equation (5.3). No unique solution exists unless we introduce a location constraint which, as indicated previously, is generally taken as $\bar{\mathbf{x}} = \mathbf{0}$ so that $\sum_{i=1}^{n} x_{ij} = 0$, for all j.

Using these constraints and equation (5.2) implies that the sum of terms in any row of $\mathbf{B}$ is zero. Consequently, summing equation (5.2) over i, over j and finally over both i and j leads to the following three equations:

$$\sum_{i=1}^{n} d_{ij}^2 = D + nb_{jj},$$

$$\sum_{j-1}^{n} d_{ij}^2 - nb_{ii} + D, \qquad (5.4)$$

$$\sum_{i=1}^{n} \sum_{j=1}^{n} d_{ij}^2 = 2nD$$

where $D = \sum_{i=1}^{n} b_{ii}$ is the trace of the matrix $\mathbf{B}$. Solving equations (5.3) and (5.4) we find that

$$b_{ij} = -\frac{1}{2}[d_{ij}^2 - d_{i.}^2 - d_{.j}^2 + d_{..}^2] \qquad (5.5)$$

where

$$d_{i.}^2 = \frac{1}{n}\sum_{i=1}^{n} d_{ij}^2, \quad d_{.j}^2 = \frac{1}{n}\sum_{i=1}^{n} d_{ij}^2, \quad d_{..}^2 = \frac{1}{n^2}\sum_{i=1}^{n}\sum_{j=1}^{n} d_{ij}^2$$

Equation (5.5) now gives us the elements of the matrix $\mathbf{B}$ in terms of squared Euclidean distances.

To factorise $\mathbf{B}$ in the form $\mathbf{XX'}$ we need to find the eigenvectors of $\mathbf{B}$ and scale them so that their sums of squares are equal to the corresponding eigenvalues. The matrix $\mathbf{X}$ is then given by

$$\mathbf{X} = [\mathbf{c}_1, \mathbf{c}_2, \ldots, \mathbf{c}_n] \qquad (5.6)$$

where $\mathbf{c}_1, \ldots, \mathbf{c}_n$ are the appropriately scaled eigenvectors of $\mathbf{B}$. (This result has been discussed previously in Section 4.4.), $\mathbf{X}$ now contains the coordinates of each point referred to the principal axes.

If we are seeking a configuration in a given number of dimensions, say p^*, then we simply examine the eigenvectors associated with the p^* largest eigenvalues. If we are unsure as to the number of dimensions to use, then we can assess the adequacy of the first p^* coordinates by the criterion, T, given by

$$T = \frac{\sum_{i=1}^{p^*} \lambda_i}{\sum_{i=1}^{n} \lambda_i} \qquad (5.7)$$

where $\lambda_1, \ldots, \lambda_n$ are the eigenvalues of $\mathbf{B}$. (Since the sum of the elements in each row of $\mathbf{B}$ is chosen to be zero, $\mathbf{B}$ will always have at least one zero eigenvalue.)

When $\mathbf{B}$ arises from *Euclidean* distances it is straightforward to show that it is positive semi-definite, has positive or zero eigenvalues, and when factored as $\mathbf{XX}'$ will lead to real (as opposed to imaginary) coordinate values. Classical scaling is, however, often applied to dissimilarity measures which are not Euclidean and where the resulting $\mathbf{B}$ matrix need not be positive semi-definite. Consequently the matrix may have a number of negative eigenvalues, and the factorisation $\mathbf{B} = \mathbf{XX}'$ may now lead to imaginary values for some of the coordinates. However, if $\mathbf{B}$ has only a *small* number of *small* negative eigenvalues, a useful coordinate representation of the dissimilarity matrix may still be obtained from the eigenvectors associated with the first few positive eigenvalues. The adequacy of the representation could now be measured by; T' given by

$$T' = \frac{\sum_{i=1}^{p^*} \lambda_i}{\sum_{i=1}^{n} |\lambda_i|} \qquad (5.8)$$

However, if $\mathbf{B}$ has a number of large negative eigenvalues, classical scaling of the dissimilarity matrix may be inadvisable, and we may prefer to use one of the methods discussed in Section 5.4.

It is easy to show that classical scaling with Euclidean distances is exactly equivalent to principal components analysis of a covariance matrix, in the sense that the coordinates produced by the former will be the same as the principal component scores of each individual. (See Chatfield and Collins 1980, and Exercise 5.1.) The equivalence of principal components analysis and classical scaling, which is often referred to as *principal coordinates analysis* (see Gower, 1966), means that there is no point in carrying out both analyses. Consequently if $n > p$, then a principal components analysis is to be preferred because it is easier to find the eigenvectors of the $(p \times p)$ matrix $\mathbf{X}'\mathbf{X}$ than those of the larger $(n \times n)$ matrix $\mathbf{XX}'$.

5.3.1   Classical Scaling — a Numerical Example

To illustrate how classical scaling might be used in practice we shall apply the method to the matrix of airline distances between the ten American cities shown in Table 5.3. These distances are not Euclidean since they relate essentially to journeys along the surface of a sphere. Table 5.4 shows the eigenvalues of the $\mathbf{B}$ matrix derived from these distances, and the eigenvalues associated with the two largest of these. Since the distance matrix is non-Euclidean, there are a number of negative eigenvalues. However, these are relatively small and Figure 5.4 confirms that the coordinates obtained from the first two eigenvectors give a very reasonable representation of the distances.

Table 5.3 Airline distances between ten US cities
(Kruskal and Wish, 1978, courtesy of Sage Publications)

	Atla	Chic	Denv	Hous	LA	Mia	NY	SF	Seat	Wash
Atlanta	—									
Chicago	587	—								
Denver	1212	920	—							
Houston	701	940	879	—						
Los Angeles	1936	1745	831	1374	—					
Miami	604	1188	1726	968	2339	—				
New York	748	713	1631	1420	2451	1092	—			
San Francisco	2139	1858	949	1645	347	2594	2571	—		
Seattle	2182	1737	1021	1891	959	2734	2408	678	—	
Washington DC	543	597	1494	1220	2300	923	205	2442	2329	—

Table 5.4 Eigenvalues and eigenvectors arising from classical
multidimensional scaling applied to distances in Table 5.3

	Eigenvalues		First two eigenvectors	
			1	2
1	9582144·3	Atlanta	718·7	−143·0
2	1686820·1	Chicago	382·0	340·8
3	8157	Denver	−481·6	25·3
4	1432·9	Houston	161 5	−527·8
5	507·7	Los Angeles	−1203·7	−390·1
6	25·1	Miami	1133·5	−501 0
7	0·0	New York	1072·2	519·0
8	−897·7	San Francisco	−1420·6	−112·6
9	−5467·6	Seattle	−1341·7	579·7
10	−35478·9	Washington DC	979·6	335·5

The criterion T' given in equation (5.8) takes the value 99% for the first two eigenvalues of Table 5.4, which again indicates the adequacy of the two-dimensional representation.

5.4 Metric and Non-metric Multidimensional Scaling

The central motivating concept of multidimensional scaling is that the distances between the points representing the items or stimuli of interest should correpond in some sensible way to the observed proximities. With this in mind various authors, for example, Shepard (1962), Kruskal (1964a) and Sammon (1969), have approached the problem by defining an objective function which measures the discrepancy between the observed proximities and the fitted distances. They then attempt to recover the configuration of points in a particular number of dimensions which minimises this function, using some type of optimisation algorithm.

For example, suppose that the proximity matrix under investigation contains a measure of dissimilarity for each pair of objects, is symmetric, and contains zeros on the main diagonal. (Its elements are represented by δ_{ij}.) What

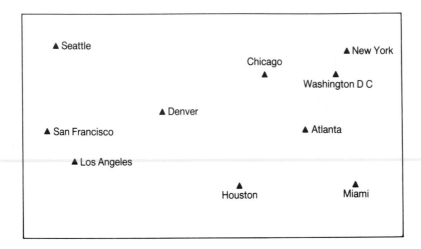

Figure 5.4 Two-dimensional solution given by classical scaling applied to the data in Table 5.3. (Kruskal and Wish, 1978)

is required is a set of d-dimensional coordinates for each object (where d is ideally 2 or 3), with associated Euclidean distances d_{ij} which match the observed dissimilarities. To assess the agreement between distances and dissimilarities we need to define a function which takes the value zero if the pattern of the distances fits that of the dissimilarities perfectly, and increases in value as the fit becomes less good. An intuitively appealing candidate for this function is a sum of squares, SS, given by

$$SS = \sum_{i=1}^{n-1} \sum_{i=i+1}^{n} (\delta_{ij} - d_{ij})^2 \qquad (5.9)$$

If the distances are all equal to the correponding dissimilarities, SS takes the value of zero. As discrepancies between d_{ij} and δ_{ij} increase so does the value of SS. Since the distances d_{ij} are a function of the n,d-dimensional coordinate values, so also is SS, and by altering the coordinate values one changes the value of SS. Since goodness-of-fit increases with decreasing values of SS, we now seek to determine that set of d-dimensional coordinates that minimises SS. Various optimisation algorithms, such as the method of steepest descent, might be considered, but the details of these need not concern us here.

The sum-of-squares criterion is invariant under rigid transformations of the configuration (rotations, translations and reflections). This is clearly desirable. Unfortunately it is not invariant under non-rigid transformations such as uniform stretching and shrinking. In other words, if we stretch the configuration $x_1, x_2, \ldots, x_n$ by the factor k to $kx_1, kx_2, \ldots, kx_n$ then SS changes. Now, purely enlarging or shrinking a configuration should not alter how well it fits the data since the relationships between the distances do not change. Consequently one should introduce a scaling factor which has the same dependence on the scale of the configuration as does SS. Such a scaling factor is easily found, for example:

$$SC = \sum_{i=1}^{n-1} \sum_{j=i+1}^{n} d_{ij}^2 \qquad (5.10)$$

A goodness-of-fit measure given by $S = SS/SC$ has all the desirable properties of SS and is, moreover, invariant under changes of scale, that is, uniform stretching or shrinking. The square root of S is generally known as *stress*.

The choice of a goodness-of-fit function based upon SS given by equation (5.9) implies that we are willing to assume that there is the following simple relationship between the observed dissimilarities and the interpoint distance

$$d_{ij} = \delta_{ij} + \epsilon_{ij} \qquad (5.11)$$

where the ϵ_{ij} represent a combination of errors of measurement and distortion errors arising because the dissimilarities may not exactly correspond to a configuration in d dimensions. It is this rather naïve assumption which is essentially the basis of classical multidimensional scaling, discussed in the previous section.

However, in many situations it may be more realistic to postulate a rather less simple relationship between the δ_{ij} and d_{ij}. For example, a straightforward extension of equation (5.11) would be

$$d_{ij} = a + b\delta_{ij} + \epsilon_{ij} \qquad (5.12)$$

The residual sum of squares criterion would now become

$$SS = \sum_{i=1}^{n-1} \sum_{j=i+1}^{n} (d_{ij} - a - b\delta_{ij})^2 \qquad (5.13)$$

Minimisation of stress would now involve a two-stage procedure. First, for a given configuration $x_1, x_2, \ldots, x_n$ and hence a given set of d_{ij} values, one would need to determine values of a and b which minimised equation (5.13). These are obtained from the simple linear regression of d_{ij} on the δ_{ij}. Now, for this value of a and b one would need to apply an optimisation algorithm to find the new values of $x_1, x_2, \ldots, x_n$ that minimise the stress criterion. These two stages would be repeated until some convergence criterion had been satisfied.

In general, if we postulate a relationship between distances and dissimilarities of the form

$$d_{ij} = f(\delta_{ij}) + \epsilon_{ij} \qquad (5.14)$$

the residual sum of squares would be

$$SS = \sum_{i=1}^{n-1} \sum_{j=i+1}^{n} (d_{ij} - f(\delta_{ij}))^2 \qquad (5.15)$$

We would now need first to minimise SS with respect to the parameters of f by some form of regression and then minimise stress with respect to the d-dimensional configuration $x_1, x_2, \ldots, x_n$

In psychological experiments the proximity matrices frequently arise from asking human subjects or observers to assess similarities or distances. These subjects, however, can usually give only ordinal judgements. For example, they can specify that one stimulus is 'larger' than another without being able to attach any value to the exact numerical difference between them. Consequently the proximity judgements do not have strict numerical significance and so one would like, if possible, to use only their ordinal properties, and find a method of multidimensional scaling whose solutions are derived using only the rank order of the proximity values. Such a method would be invariant under monotonic transformations of the proximity matrix. A major breakthrough in multidimensional scaling was achieved in the early 1960s by Shepard (1962) and Kruskal (1964a) when they derived a method which met these requirements by allowing the function f in equation (5.14) to indicate a *monotonic relationship* between distance and dissimilarity. In other words, the distances are now represented as follows:

$$d_{ij} = \hat{d}_{ij} + \epsilon_{ij} \qquad (5.16)$$

where the $\hat{d}_{ij}$ are a set of numbers which are monotonic with the δ_{ij}. That is, if the observed dissimilarities are ranked from lowest to highest as

$$\delta_{i_1 j_1} < \delta_{i_2 j_2} < \cdots < \delta_{i_N j_N}$$

where $N = n(n-1)/2$, then the $\hat{d}_{ij}$ will satisfy

$$\hat{d}_{i_1 j_1} \le \hat{d}_{i_2 j_2} \le \cdots \le \hat{d}_{i_N j_N}$$

Stress now becomes

$$\text{Stress} = \sqrt{\left[\frac{\left[\sum_{i=1}^{n-1} \sum_{j=i+1}^{n} (d_{ij} - \hat{d}_{ij})^2\right]}{\left[\sum_{i=1}^{n-1} \sum_{j=i+1}^{n} d_{ij}^2\right]} \right]} \qquad (5.17)$$

In this case the fitted $\hat{d}_{ij}$ values are obtained from a special type of regression of distance on dissimilarities, known as *monotonic regression*. Details of the technique need not concern us here; they are available in Kruskal (1964b). The important point to emphasise is that the observed dissimilarities now only enter the calculations in terms of their rank order.

As we have seen, minimisation of the stress goodness-of-fit criterion involves two stages. The first is some form of regression analysis, which is straightforward. The second entails the minimisation of a function of nd variables (the d dimensional coordinate of each stimulus). Various optimisation algorithms have been used in this stage (for example, steepest descent and Newton–Raphson) and technical details can be found in Kruskal (1964b) and Ramsay (1977). Essentially, however, all such algorithms start with an arbitrary set of coordinate values for each object, move all the points a little to achieve a lower stress and then repeat the procedures until a configuration is reached from which no improvement is possible. Roughly speaking, points x_i and x_j are moved closer together if $f(\delta_{ij}) < d_{ij}$ and apart in the opposite case, making d_{ij} more like $f(\delta_{ij})$. Since the stress function may have a number of minima, problems can arise from convergence to a *local* rather than to the *global* minimum. This problem is unavoidable, but can be partially overcome by repeating the calculations with different initial configurations.

5.5   Choosing the Number of Dimensions

The decision about the number of coordinates needed for a given data set is as much a substantive question as a statistical one. Even if a reasonable statistical method existed for determining the 'correct' or 'true' dimensionality, this would not in itself be sufficient to indicate how many coordinates the researcher needs to use. Since multidimensional scaling is almost always used as a descriptive model for representing and understanding a data set, other considerations enter into decisions about the appropriate dimensionality. This point is made by Gnanadesikan and Wilk (1969):

> Interpretability and simplicity are important in data analysis and any rigid inference of optimal dimensionality, in the light of the observed values of a numerical index of goodness-of-fit, may not be productive.

In the light of such comments, two-dimensional solutions are likely to be those of most practical importance since they have the virtue of simplicity, are often readily assimilated by the investigator, and may, in many cases, provide an easily understood basis for the discussion of observed proximity matrices. Nevertheless there may be occasions when two dimensions are just not adequate to contain the full complexity of the structure present and the investigator would like some guidance on a reasonable number of coordinates to use to represent the data. Perhaps the most commonly used procedure seeking to give such guidance is that based upon examining stress values for different numbers of dimensions. In his original paper Kruskal (1964a) gave the following advice about stress, based upon his experience with experimental and synthetic data.

Stress (%)	*Goodness-of-fit*
20	poor
10	fair
5	good
2.5	excellent
0	perfect

Consequently, observed stress values might be 'evaluated' against these comments as an indication of when the fit for a particular number of dimensions is 'good' or better. In addition, it has been suggested that the stress may be plotted against the number of dimensions, and the diagram examined for the presence of an 'elbow' indicating the appropriate number of coordinates to use. Examples of such plots are given in the next section. However, Wagenaar and Padmos (1971) indicate that the interpretation of stress is strongly dependent on the number of stimuli involved, and that a simple interpretation in terms of Kruskal's verbal evaluation is often not justified. Spence (1970 and 1972) and Spence and Graef (1974) have carried out an extensive set of Monte Carlo experiments, the results of which allow a more objective assessment of underlying dimensionality to be made. The simulated data were generated for a wide range of conditions similar to those that might be experienced by typical users. The number of points was varied from 12 to 36, spaces of true dimensionality from one to four were investigated, and the level of error in the data varied from zero to an infinite amount. An attempt was then made to find the set of simulated values, for some given dimensionality, which best fitted the observed stress values for the data set obtained by the application of multidimensional scaling for different numbers of dimensions. The procedure is described in detail in Spence and Graef (1974).

Table 5.5 Dissimilarity matrix arising from minor distortions of the Euclidean distances between the points in Figure 5.5

	1	2	3	4	5	6	7	8	9
1	0·0								
2	2·0	0·0							
3	4·0	2·0	0·0						
4	2·0	3·0	4·2	0·0					
5	2·8	2·0	2·9	2·1	0·0				
7	4·5	3·4	2·1	4·3	2·1	0·0			
7	4·0	4·5	5·5	2·0	3·0	4·2	0·0		
8	4·2	4·0	4·3	3·0	2·1	3·1	2·1	0·0	
9	6·0	4·5	4·1	4·5	2·5	2·1	4·1	2·1	0·0

5.6 Further Examples of the Application of Multidimensional Scaling Techniques

We will begin by examining in some detail the analysis of the proximity matrix shown in Table 5.5, which arises from minor distortions of the Euclidean distances between the points in Figure 5.5. Here there will be essentially a simple linear relationship between derived distances and dissimilarities, and the analysis of the proximity matrix by the procedures outlined in Section 5.4, with the *assumption* that the function f is linear, results in the two-dimensional solution shown in Figure 5.6. As one would expect, the configuration of Figure 5.5 has been completely recovered by this analysis (apart from the orientation of axes, which is, of course, arbitrary). A scatterplot of distance against dissimilarities (Figure 5.7) confirms that our assumption of a linear relationship for these data is reasonable.

The stress for the two-dimensional solution is 12.93% indicating only a 'fair' fit according to Kruskal's ratings. However, the plot of stress against number of dimensions (Figure 5.8) clearly indicates that a two-dimensional solution is appropriate.

Now we shall transform the dissimilarities of Table 5.5 as follows:

$$\delta'_{ij} = \exp(\delta_{ij}) + \delta_{ij}^2 \qquad (5.18)$$

Application of the method of multidimensional scaling used above now results in the solution shown in Figure 5.9, which, apart from minor distortions, again recovers the input configuration. In part this is somewhat surprising since the assumption that f is linear is now quite unrealistic. However, this example serves to indicate that metric multidimensional scaling is robust, in the sense that the configuration obtained using one assumption about f will not, in general, be very different from that using an alternative assumption, although, as with this example, the stress may be highly inflated by the inappropriate assumption made about f. (The stress for the configuration in Figure 5.9 is 54.8%.) Examination of the scatterplot of distances against dissimilarities in this case (see Figure 5.10) indicates that the assumption of a linear relationship is wrong. In such cases it is best to reanalyse the data using a more appropriate assumption. Application of non-metric multidimensional scaling to either the

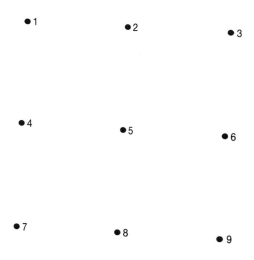

Figure 5.5 Artificial data set

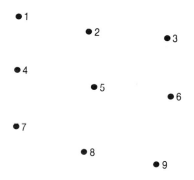

Figure 5.6 Two-dimensional solution obtained from multidimensional scaling assuming a linear relationship between distances and dissimilarities applied to the data in Table 5.5

original or the transformed dissimilarities results in the same solution with the same stress (2.12%) since δ'_{ij} is simply a monotonic transformation of δ_{ij}. This serves to indicate the possible advantages of this method over those specifying a more rigid relationship between distances and dissimilarities.

As a further example we will analyse the similarity matrix shown in Table 5.6, which arises from an investigation into the assessment of pain. The entries in Table 5.6 are the similarities for pairs of adjectives describing pain, derived by a method suggested by Burton (1972). The two-dimensional solution is shown in Figure 5.11. (Two dimensions was suggested by use of the technique of Spence and Graef; see Section 5.5). The detailed analysis of these data is described in Reading, Everitt and Sledmere (1982), but one interpretation,

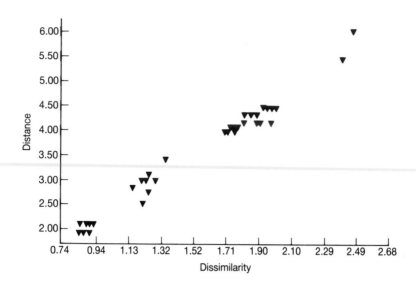

Figure 5.7   Scatterplot of distances against dissimilarities for the example in Figure 5.6

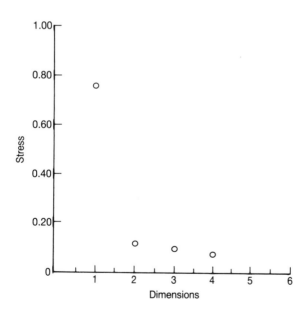

Figure 5.8   Plot of number of dimensions against stress for multidimensional scaling of artificial data set of Figure 5.5

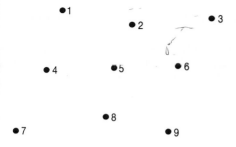

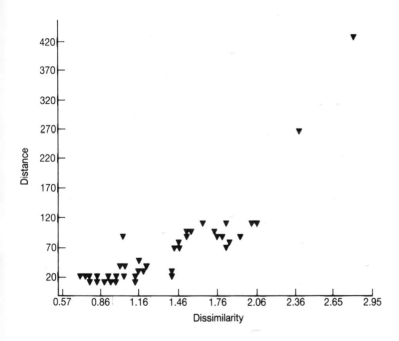

Figure 5.9 Two-dimensional solution obtained from multidimensional scaling assuming a linear relationship between distances and dissimilarities applied to the transformed dissimilarities in Table 5.5

Figure 5.10 Scatterplot of distances against dissimilarities for the example in Figure 5.9

obtained by a rotation of the axes, is indicated in Figure 5.11. One axis now ranges from words such as 'tingling', 'quivering' and 'pricking' to words such as 'wrenching', 'crushing' and 'pressing', and might be suggestive of a dimension ranging from non-intrusive to intrusive pain. The second axis has adjectives such as 'throbbing', 'beating' and 'pulsing' at one extreme and 'sharp', 'cutting' and 'lancinating' at the other, and could perhaps be a dimension representing non-specific to specific pain. Of course, such interpretations are relatively subjective, and readers might well come up with others which they find more plausible.

The search for meaningful dimensions is often one of the main aims of scaling and Kruskal and Wish (1978) suggest that multiple linear regression between some direction in the configuration and various characteristics of the points may be useful in trying to interpret the dimensions.

5.7   Non-Euclidean Metrics

Up to now we have considered only the use of Euclidean distances in multi-dimensional scaling procedures. In principle, however, there is no reason why the goodness-of-fit function, stress, could not be used with almost any kind of distance. For example, Kruskal (1964a) has considered a class of distances known as *Minkowski metrics* which are defined as follows:

$$d_{ij} = \left[\sum_{k=1}^{d} |x_{ik} - x_{jk}|^r \right]^{\frac{1}{r}} \tag{5.19}$$

When $r = 1$ we have the fairly well-known *city block metric;* and when $r = 2$, equation (5.19) reduces to the Euclidean distance. Kruskal (1964a) gives an example of using such metrics on a set of experimental data obtained by Ekman (1954), from the judgement of similarities of fourteen colour stimuli. Figure 5.12 shows the different values of stress obtained for different values of r in equation (5.19), for a two-dimensional solution. From this we see that a value of $r = 2.5$ gives the best fit, indicating perhaps, that subjective distance between colours may be slightly non-Euclidean. The actual two-dimensional solution obtained in each case was very similar in general terms, although the precise shape, spacing and angular orientation varied with r.

The Minkowski metrics differ sharply from Euclidean distance in not being invariant under rigid rotation of the coordinate axes. Thus, while a configuration may be freely rotated when Euclidean distances are being used, it may not be when more general distances are employed.

In most applications of multidimensional scaling, Euclidean distance is likely to be of greatest use, but non-Euclidean distances such as those specified in equation (5.19) may be worth considering in some situations.

5.8   Three-way Multidimensional Scaling

The methods of multidimensional scaling discussed in Sections 5.3 and 5.4 are designed essentially for the analysis of *two-way* matrices of proximities. In many situations the investigator may have several such matrices for the same set of objects, for example, one from each subject, and these may be thought of as forming *a three-way matrix* (see Figure 5.13). A method such as non-metric multidimensional scaling could be used to analyse such data by simply averaging the proximities over subjects (see, for example, the similarity of nations example in Section 5.2). Such an approach, however, implicitly assumes that differences between subjects are simply due to random error.

Consequently we would prefer a method capable of dealing with such three-way data which allows for the possibility of large systematic differences among

Table 5.6   Similarities of pairs of adjectives describing pain

	1	2	3	4	5	6	7	8	9	10	11	12	13	14	15	16	17	18	19	20	21	22	23	24	25	26	27	28	29
2	106·36																												
3	27·72	30·33																											
4	19·45	16·23	144·52																										
5	27·04	13·21	102·95	113·04																									
6	25·06	21·85	130·16	141·26	118·70																								
7	96·13	80·49	31·05	23·00	23·60	23·04																							
8	72·58	49·70	23·71	16·14	17·08	17·64	73·81																						
9	40·17	28·41	16·51	9·39	14·57	12·41	60·16	70·63																					
10	28·03	17·77	−3·61	−5·42	−4·98	−6·94	8·98	9·14	3·27																				
11	−11·40	−5·44	−11·40	−10·10	−10·10	−10·10	−5·88	−11·40	7·50	14·47																			
12	−9·67	−5·48	−1·22	−4·58	−4·58	1·07	−4·46	−2·87	26·46	20·73	100·89																		
13	−3·26	−1·64	3·96	17·63	10·36	11·51	−0·74	2·85	40·82	44·64	23·44	47·08																	
14	−5·44	−7·57	−5·69	−8·29	−4·36	−6·14	−2·14	−3·36	3·44	22·37	15·38	21·54	45·75																
15	−10·29	−10·29	−11·40	−11·40	−6·67	−10·25	−2·49	3·71	10·64	31·03	−4·37	5·84	50·46	39·87															
16	−9·82	−8·95	−9·82	−11·40	−9·43	−10·25	−6·87	−4·71	7·00	2·63	0·57	15·51	34·72	35·16	65·25														
17	−11·40	−11·40	−9·40	−9·40	−5·47	−8·25	−5·85	−3·32	6·89	7·82	−2·47	6·79	27·84	43·73	34·05	118·77													
18	−0·96	−2·17	−7·29	−6·95	−2·10	−6·73	0·98	1·34	−5·42	15·97	−6·99	−2·11	−5·60	−2·67	9·03	4·76	−0·98												
19	−9·27	−3·95	−2·69	−4·14	4·50	0·38	−9·09	−7·30	−4·70	−6·75	−3·97	6·41	−4·48	−2·34	−4·16	−7·30	−4·44	33·09											
20	−7·25	−1·49	−2·67	−0·74	−1·18	−3·30	−6·98	−1·96	−5·40	−5·36	11·01	5·29	−7·14	−7·98	−7·37	−8·62	1·42	−2·37											
21	−2·80	−5·72	−4·02	−3·59	6·92	−1·21	−1·12	−4·82	−4·06	−7·76	−1·48	−2·97	−7·44	−7·07	−4·81	−6·28	−8·12	22·35	21·76	13·61									
22	9·24	8·23	−4·19	−6·98	−0·22	1·00	−7·48	−8·33	−3·95	−11·40	−4·72	3·86	−6·61	−6·99	−7·48	−8·56	−2·96	23·24	117·46	−8·55	11·56								
23	−5·91	−9·72	−0·18	−1·67	1·30	0·68	0·15	−11·40	−6·89	−7·92	−5·56	0·73	−9·53	−7·03	−9·60	−2·96	−5·16	11·90	−1·39	6·88	6·01	2·35							
24	−11·40	−10·26	−3·75	0·40	−1·66	−2·90	−3·42	−7·30	8·73	−10·07	−9·72	−1·29	−8·44	−7·71	−8·55	−5·91	−7·33	16·25	5·45	13·92	10·18	1·19	135·22						
25	−9·87	−8·74	6·98	−6·71	−6·52	−8·23	−8·50	−6·76	−2·39	−9·87	−1·76	7·97	−9·37	−5·64	−9·20	−0·02	7·25	10·76	6·66	−3·89	36·50	11·92	65·80	63·89					
26	−9·94	−8·34	−9·93	−10·10	−9·31	−10·10	−9·93	−6·91	−11·40	−8·50	−10·10	−11·40	−11·40	−11·40	−7·41	−11·40	−9·96	−9·15	−8·35	−11·40	−11·40	−11·40	−11·40	−11·40	−11·40				
27	−9·67	−9·67	−11·40	−11·40	−10·61	−11·40	−11·40	−6·97	−6·94	−10·01	−11·40	−9·67	−4·47	−11·40	−6·12	−10·06	−8·67	−9·21	−10·10	−11·40	−7·01	−9·67	−8·71	−10·10	−8·71	162·96			
28	−9·67	−8·07	−11·40	−11·40	−10·61	−11·40	−8·31	−6·85	−3·85	−10·20	−11·40	−8·48	−4·24	−10·20	−4·80	−7·35	−5·96	−9·41	−8·28	−11·40	−11·40	−9·67	−10·00	−9·89	−10·00	170·71	178·41		
29	−9·02	−9·02	−8·18	−6·88	−4·12	−5·73	−5·25	2·33	13·99	−2·37	−4·86	3·25	7·53	7·72	6·09	25·14	41·90	−9·21	−10·03	8·76	−11·40	−7·21	−6·26	−4·86	9·70	29·66	39·74	38·03	
30	23·12	25·19	−2·31	−5·58	−3·62	−5·58	6·90	17·71	2·81	54·48	−5·76	−7·09	−1·31	−4·20	0·21	−7·19	−4·11	7·33	−8·52	−4·96	−2·56	−9·96	−10·07	−10·07	−11·40	−4·35	−4·22	−7·51	−7·06

Key

1 Flickering	7 Jumping	13 Stabbing	19 Pressing	25 Wrenching
2 Quivering	8 Flashing	14 Lancinating	20 Gnawing	26 Hot
3 Pulsing	9 Shooting	15 Sharp	21 Cramping	27 Burning
4 Throbbing	10 Pricking	16 Cutting	22 Crushing	28 Scalding
5 Beating	11 Boring	17 Lacerating	23 Tugging	29 Searing
6 Pounding	12 Drilling	18 Pinching	24 Pulling	30 Tingling

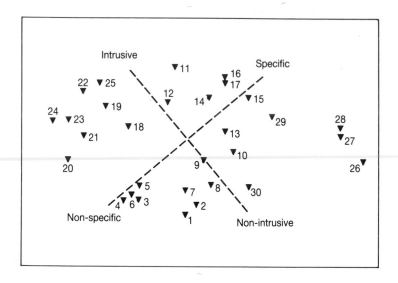

Figure 5.11   Two-dimensional solution obtained from non-metric scaling of the pain adjectives similarity matrix in Table 5.6

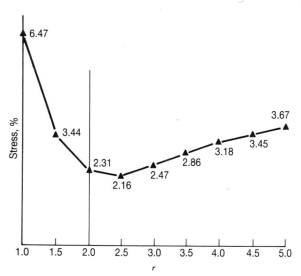

Figure 5.12   Plot of stress against value of *r* in Minkowski metrics formula for colour stimulus data. (Taken from Kruskal and Wish, 1978, courtesy of Sage Publications)

the observed proximity matrices of different subjects. Several methods have been suggested by, amongst others, Tucker (1964 and 1972), Harshman (1972)

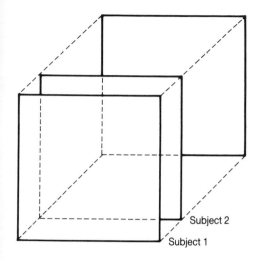

Subject 2

Subject 1

Figure 5.13 Three-way data matrix

and Tucker and Messick (1963). However, the most successful approach would seem to be the INDSCAL model, proposed by Carroll and Chang (1970), which is the focus of our attention here. (INDSCAL stands for individual scaling or individual differences scaling.)

This model assumes that there is a set of d dimensions *common* to all subjects in which the n objects or stimuli may be represented, but that the distance between points in this space differ from subject to subject according to the importance, or *weight*, attached to each dimension by a particular subject. So the 'modified' Euclidean distance between points i and j for the lth subject is given by

$$d_{ij}^{(l)} = \left[\sum_{k=1}^{d} w_{lk}(x_{ik} - x_{jk})^2 \right]^{\frac{1}{2}} \tag{5.20}$$

where the weights $w_{lk}, k = 1, \ldots, d$ represent the differing importance attached to each dimension by subject l. Another way of looking at formula (5.20) is to say that the $d_{ij}^{(l)}$ are ordinary Euclidean distances computed in a space whose coordinates are

$$y_{ik}^{(l)} = w_{lk}^{\frac{1}{2}} x_{ik}, \tag{5.21}$$

that is, a space like the x-space except that the configuration has been expanded or contracted (differentially) in directions corresponding to the coordinate axes. The model is illustrated in Figure 5.14. An important point to note about the INDSCAL model is that it produces a *unique* orientation of the axes of the space in which the stimuli or objects are represented as points. This arises because subjects are permitted to stretch or shrink axes differentially, whereas only a uniform stretching preserves distances. Consequently it is not permissible to rotate the axes here as it was for the methods of scaling described

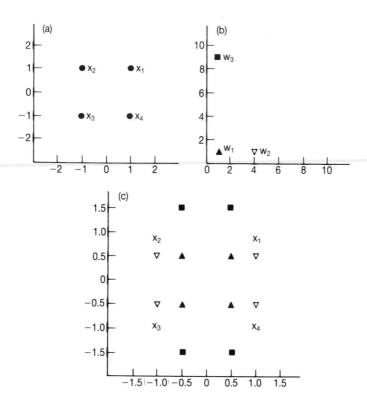

Figure 5.14   Illustration of the INDSCAL model: (a) common stimulus space; (b) subject weights for two-dimensions; (c) 'private' stimulus space for the three individuals with weights as in (b). (Adapted from Kruskal and Wish, 1978, courtesy of Sage Publications)

in Sections 5.3 and 5.4. Fortunately it has usually been found that the recovered dimensions are fairly readily interpretable. A linear relationship is assumed between the distances given by equation (5.20) and the observed dissimilarities, and the first step in fitting this model is to convert the dissimilarity matrix of each subject into a corresponding **B** matrix, using equation (5.5). Least squares estimates of the required weights and coordinates are then found by a procedure which Carroll and Chang call the *canonical decompositon* of N-way tables. This involves an iterative procedure in which, for given initial estimates of the coordinates, least squares estimates of the weights are determined. These are then used to find revised least squares estimates of the coordinates and the procedure is repeated until some convergence criterion is satisfied. For details the reader is referred to the original paper by Carroll and Chang (1970).

We will now examine an application of this technique to some actual data, arising from dissimilarity ratings made by a number of subjects on politicians prominent at the time of the Second World War. The ratings were made on a nine-point scale from 1, indicating very similar, to 9, indicating very dissimilar. Two of the dissimilarity matrices collected are shown in Table 5.7.

Table 5.7 Dissimilarity ratings of World War II politicians by two subjects: subject 1 in lower triangle; subject 2 in upper triangle

		1	2	3	4	5	6	7	8	9	10	11	12
1	*Hitler*		2	7	8	5	9	2	6	8	8	8	9
2	*Mussolini*	3		8	8	8	9	1	7	9	9	9	9
3	*Churchill*	4	6		3	5	8	7	2	8	3	5	6
4	*Eisenhower*	7	8	4		8	7	7	3	8	2	3	8
5	*Stalin*	3	5	6	8		7	7	5	6	7	9	5
6	*Attlee*	8	9	3	9	8		9	7	7	4	7	5
7	*De Gaulle*	4	4	3	5	6	5		4	6	5	6	5
8	*Mao Tse Tung*	8	9	8	9	6	9	8		7	8	8	6
10	*Truman*	9	9	5	4	7	8	8	4	4		4	6
11	*Chamberlain*	4	5	5	4	7	2	2	5	9	5		8
12	*Tito*	7	8	2	4	7	8	3	2	4	5	7	

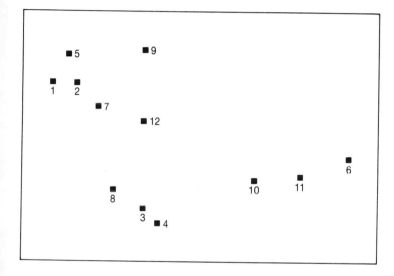

Figure 5.15 Two-dimensional solution obtaines from INDSCAL applied to dissimilarity ratings of World War II politicians

The two-dimensional INDSCAL solution obtained from the dissimilarity matrices of eight subjects is shown in Figure 5.15. One possible interpretation of the two dimensions is that the first is some indication of the 'historical impact' of the politician ranging from Attlee at one end to Hitler at the other. The second might tentatively be labelled as a 'democracy' dimension, ranging from Eisenhower, Churchill and De Gaulle through Tito, to Mao Tse Tung.

In addition to the diagram in Figure 5.15, INDSCAL produces a set of weights for each subject on each dimension. With a two-dimensional solution the pairs of weights for each subject may be further plotted to give the 'subject space'.

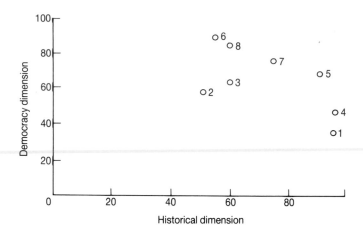

Figure 5.16 Subject weights from INDSCAL

This is shown in Figure 5.16. From this we can see that, for example, subject 1 weights the historical impact dimension rather more than the democracy dimension, whilst for subjects 6 and 8 the reverse situation holds. It is often useful to relate these weights to other characteristics of the subjects. For example, if some biographical or attitudinal information is available, we could investigate whether subjects having high weight on a dimension differ from those with low weight on any of the background variables. By applying a subject's weights to the respective dimensions of Figure 5.15, one obtains the 'private' stimulus space of that subject. The private spaces of subjects 1 and 6 are shown in Figure 5.17. This example demonstrates how the INDSCAL model can accommodate large differences between individuals, and although the model is likely to be perhaps an oversimplification for many data sets, it can often still prove extremely useful in identifying and characterising important variation in proximity data.

5.9   The Analysis of Asymmetric Proximity Matrices

In most applications of multidimensional scaling the observed proximity matrix is symmetric, that is

$$\delta_{ij} = \delta_{ji} \qquad (5.22)$$

However, situations can arise in which there are asymmetric relationships between pairs of objects so that the equality (5.22) no longer holds. Take, for example, the data described in Section 5.2 concerning Morse code signals, which is shown in its complete form in Table 5.8. This table is not symmetric since the percentage of times that signal i followed by signal j was said to be the same is not necessarily the same as when the signals are presented in the reverse order. Constantine and Gower (1978) suggest a number of other applications where assymetry may arise, including diallel cross experiments giving the number of progeny or yield when a male of line i is crossed with

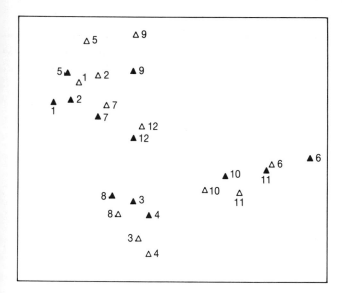

Figure 5.17   'Private' spaces of individuals 1 and 6

a female of line j, the number of people who live in location i and work in location j and the frequency with which journal i cites journal j. Such asymmetric matrices are frequently made suitable for analysis by one of the methods described previously by the transformation

$$\delta'_{ij} = \frac{1}{2}(\delta_{ij} + \delta_{ji}) \qquad (5.23)$$

However, this ignores departures from symmetry which may be informative, and Gower (1977) suggests a number of other methods which may be used directly on asymmetric matrices to obtain a spatial representation. Here we shall concentrate on just one of these, namely *multidimensional unfolding,* which derives from the 'unfolding' method of Coombs (1964), originally designed to scale preference data.

To introduce the method suppose that one has a number of subjects, say n, who are asked to rank a number of stimuli, say m, in order of preference. So the data matrix consists of n rows each containing an ordering of the m stimuli. (Note that there is no need for n to equal m.) Multidimensional unfolding attempts to find a set of coordinates for the subjects and a set of coordinates for the stimuli such that the ordering of the Euclidean distances between a given individual and each of the stimuli corresponds to the observed preference order for that individual in the sense that small distances will correspond to the most preferred stimuli.

The problem of finding the coordinates is approached in a manner very similar to non-metric multidimensional scaling as described in Section 5.4. In this case, however, the goodness-of-fit criterion to be minimised takes a

Table 5.8   Data matrix of percentages of confusions between pairs of Morse code signals (Rothkopf, 1957; Shepard, 1962)

Second signal

	A	B	C	D	E	F	G	H	I	J	K	L	M	N	O	P	Q	R	S	T	U	V	W	X	Y	Z	1	2	3	4	5	6	7	8	9	0	
A	92	04	06	13	03	14	10	13	46	05	22	03	25	34	06	06	09	35	23	06	37	13	17	12	07	03	02	07	05	05	08	06	05	06	02	03	A
B	05	84	37	31	05	28	17	21	05	19	34	40	06	10	12	22	25	16	18	02	18	34	08	84	30	42	12	17	14	40	32	74	43	17	04	04	B
C	04	38	87	17	04	29	13	07	11	19	24	35	14	03	09	51	34	24	14	06	06	11	14	32	82	38	13	15	31	14	10	30	28	24	18	12	C
D	08	62	17	88	07	23	40	36	09	13	81	56	08	07	09	27	09	45	29	06	17	20	27	40	15	33	03	09	06	11	09	19	08	10	05	05	D
E	06	13	14	06	97	02	04	04	17	01	05	06	04	04	05	01	05	10	07	67	03	03	02	05	06	05	04	03	05	03	05	02	04	02	03	03	E
F	04	51	33	19	02	90	10	29	05	33	16	50	07	06	10	42	12	35	14	02	21	27	25	19	27	13	08	16	47	25	26	24	21	05	05	05	F
G	09	18	27	38	01	14	90	06	05	22	33	16	14	13	62	52	23	21	05	03	15	14	32	21	28	14	05	10	04	10	17	23	20	11			G
H	03	45	23	25	09	32	08	87	10	10	09	29	05	08	08	14	08	17	37	04	36	59	09	33	14	11	03	09	15	43	70	35	17	04	03	03	H
I	64	07	07	13	10	08	06	12	93	03	05	16	13	30	07	03	05	19	35	16	10	05	08	02	05	07	02	05	08	09	06	08	05	02	04	05	I
J	07	09	38	09	02	24	18	05	04	85	22	31	08	03	21	63	47	11	02	07	09	09	09	22	32	28	67	66	33	15	07	11	28	29	26	23	J
K	05	24	38	73	01	17	25	11	05	27	91	33	10	12	31	14	31	22	02	02	23	17	33	63	16	18	05	09	17	08	08	18	14	13	05	06	K
L	02	69	43	45	10	24	12	26	09	30	27	86	06	02	09	37	36	28	12	05	16	19	20	31	25	59	12	13	17	15	26	29	36	16	07	03	L
M	24	12	05	14	07	17	29	08	08	11	23	08	96	62	11	10	15	20	07	09	13	04	21	09	18	08	05	07	06	06	05	07	11	07	10	04	M
N	31	04	13	30	08	12	10	16	13	03	16	08	59	93	05	09	05	28	12	10	16	04	12	04	06	11	05	02	03	04	04	06	02	02	10	02	N
O	07	07	20	06	05	09	76	07	02	39	26	10	04	08	86	37	35	10	03	04	11	14	25	35	27	27	19	17	07	07	06	18	14	11	20	12	O
P	05	22	33	12	05	36	22	12	03	78	14	46	05	06	21	83	43	23	09	04	12	19	19	19	41	30	34	44	24	11	15	17	24	23	25	13	P
Q	08	20	38	11	04	15	10	05	02	27	83	26	07	06	22	51	91	11	02	03	06	14	12	37	50	63	34	32	17	12	09	27	40	58	37	24	Q
R	13	14	16	23	05	34	26	15	07	12	21	33	14	12	12	29	08	87	16	02	23	23	62	14	12	13	07	10	13	04	07	12	07	09	01	02	R
S	17	24	05	30	11	26	05	59	16	03	13	10	05	17	06	06	03	18	96	09	56	24	12	10	06	03	03	03	08	07	06	14	06				S
T	05	13	10	01	05	46	03	06	06	14	06	14	07	06	05	06	11	04	04	07	96	08	05	04	02	02	06	05	03	03	08	07	06	14	06		T
U	14	29	12	32	04	32	11	34	21	07	44	32	11	13	06	20	12	40	51	06	93	57	34	17	09	11	06	06	16	34	10	09	09	07	04	03	U
V	05	17	24	16	09	29	06	39	05	11	26	43	04	01	09	17	10	17	11	06	32	92	17	57	35	10	10	14	28	79	44	36	25	10	01	05	V
W	09	21	30	22	09	36	25	15	04	25	29	18	15	06	26	20	25	61	12	04	19	20	86	22	25	22	10	12	29	19	16	05	09	11	06	03	W
X	07	64	45	19	03	28	11	06	01	35	50	42	10	08	24	32	61	10	12	03	12	17	21	91	48	26	12	20	24	27	16	57	29	16	17	06	X
Y	09	23	04	26	22	09	01	30	12	14	05	06	14	30	52	05	07	04	06	13	21	44	86	23	26	44	40	15	11	26	22	33	23	16			Y
Z	03	46	45	18	02	22	17	10	07	23	21	51	11	02	15	59	72	14	04	03	09	11	12	36	42	87	16	21	27	09	10	25	66	47	15	15	Z
1	02	05	10	03	03	05	13	04	02	29	05	14	09	07	14	30	28	09	04	02	03	12	14	17	19	22	84	63	13	08	10	08	19	32	57	55	1
2	07	14	22	05	04	20	13	03	25	26	09	14	02	03	17	37	28	06	05	03	06	10	11	17	30	13	62	89	54	20	05	14	20	21	16	11	2
3	03	08	21	05	04	32	06	12	02	23	06	13	05	02	05	37	19	09	07	06	04	16	06	22	25	12	18	64	86	31	23	41	16	17	08	10	3
4	06	19	19	12	08	25	14	16	07	21	13	19	03	03	02	17	29	11	09	03	17	55	08	37	24	03	05	26	44	89	42	44	32	10	03	03	4
5	08	45	15	14	02	45	04	67	07	14	04	41	02	00	04	13	07	09	27	02	14	45	07	45	10	14	10	30	69	90	42	24	10	06	05	05	5
6	07	80	30	17	04	23	04	14	02	11	11	27	06	02	07	16	30	11	14	03	12	30	09	58	38	39	15	14	26	24	17	88	69	14	05	14	6
7	06	33	22	14	05	25	06	04	06	24	13	32	07	06	07	36	39	12	06	02	03	13	09	30	30	50	22	29	18	15	12	61	85	70	20	13	7
8	03	23	40	06	03	15	15	06	02	33	10	14	03	06	14	12	45	02	06	04	06	07	05	24	35	50	42	29	16	16	09	30	60	89	61	26	8
9	03	14	23	03	01	06	14	05	02	30	06	07	16	11	10	31	32	05	06	07	06	03	08	11	21	24	57	39	09	12	04	11	42	56	91	78	9
0	09	03	11	02	05	07	14	04	05	30	08	03	02	03	25	21	29	02	03	04	05	03	02	12	15	20	50	26	09	11	05	22	17	52	81	94	0

First signal (label at left of the matrix)

	A	B	C	D	E	F	G	H	I	J	K	L	M	N	O	P	Q	R	S	T	U	V	W	X	Y	Z	1	2	3	4	5	6	7	8	9	0

different form, namely

$$S_2 = \sqrt{\frac{\left[\frac{1}{n}\sum_{i=1}^{n}\sum_{j=1}^{m}(d_{ij}-\hat{d}_{ij})^2\right]}{\sum_{j=1}^{m}(d_{ij}-\overline{d}_i)^2}} \qquad (5.24)$$

where d_{ij} is the Euclidean distance between the point representing subject i and that representing stimuli j, $\overline{d}_i = \frac{1}{m}\sum_{j=1}^{m} d_{ij}$ is the mean distance of subject i from the m stimuli; and the numbers $\hat{d}_{ij}$ are such that $\hat{d}_{ij} < \hat{d}_{ik}$ when subject i prefers stimuli j over stimuli k. In other words the $\hat{d}_{ij}$ will have the same rank order as does subject i's preferences for the m stimuli. S_2 is again seen to be essentially a sum-of-squares criterion, but now involving only those squared deviations relevant to each subject.

When the data matrix is square $(n = m)$, with its rows and columns similarly classified (as with the Morse signal data), multidimensional unfolding will lead to n points representing the row objects and n points representing the columns.

Table 5.9 Numbers of persons (%) claiming to speak a language 'enough to make yourself understood' (adapted from Hartigan, 1975)

	Country	*Language*											
		1 Ger	*2* Ital	*3* Fr	*4* Dch	*5* Flem	*6* GB	*7* Port	*8* Swed	*9* Dan	*10* Nor	*11* Finn	*12* Spn
1	West Germany	100	2	10	2	1	21	0	0	0	0	0	1
2	Italy	3	100	11	0	0	5	0	0	0	0	0	1
3	France	7	12	100	1	1	10	1	2	3	0	0	7
4	Netherlands	47	2	16	100	100	41	0	0	0	0	0	2
5	Belgium	15	2	44	0	59	14	0	0	0	0	0	1
6	Great Britain	7	3	15	0	0	100	0	0	0	0	0	2
7	Portugal	0	1	10	0	0	9	100	0	0	0	0	2
8	Sweden	25	1	6	0	0	43	0	100	10	11	5	1
9	Denmark	36	3	10	1	1	38	0	22	100	20	0	1
10	Norway	19	1	4	0	0	34	1	25	19	100	0	0
11	Finland	11	1	2	0	0	12	0	23	0	0	100	0
12	Spain	1	2	11	0	0	5	0	0	0	0	0	100

Interpretation of the data lies mainly in investigating the distances *between* the two sets, and there is only secondary interest in the within-set distances.

We will now examine an example of the application of this technique using the data shown in Table 5.9, taken from Hartigan (1975). The percentages given in this table may be thought of as defining a rank order of 'preferences' of a country for a particular language. The two-dimensional solution obtained from multidimensional unfolding appears in Figure 5.18. The main interest in this diagram lies in the distances apart of points representing countries and those representing languages. Of particular note are the fairly marked discrepancies between the points representing Sweden and Swedish and Finland and Finnish, this reflecting the fact that a large number of Swedes and Finns speak languages other than their own, whereas there are few nationals of other countries who speak either Swedish or Finnish.

5.10 A Statistical Approach to Multidimensional Scaling

Ramsay (1977, 1978, 1982), describes a method of multidimensional scaling which includes a stochastic component and the application of maximum likelihood techniques to the estimation of coordinates. Such an approach allows confidence regions to be found for the points in a solution configuration. The essential features of Ramsay's model are as follows:-

(1) Observed dissimilarities d_{ijr} for stimuli i and j and subject r.
(2) Fitted distances of the form

$$d^*_{ijr} = (\mathbf{x}_i - \mathbf{x}_j)'\mathbf{W}_r(\mathbf{x}_i - \mathbf{x}_j) \tag{5.25}$$

The matrix $\mathbf{W}_r$ is allowed to vary from subject to subject; usual $\mathbf{W}_r$ is taken to be diagonal, so that the distance model used is the same as in INDSCAL (see previous section).

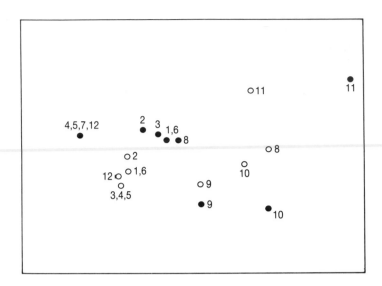

Figure 5.18   Two-dimensional solution from multidimensional unfolding of the language data in Table 5.9

(3)   Computes distances (and hence coordinates) so as to provide optimal fit of $\ln d_{ijr}^{*}$ to $f_r(\ln d_{ijr})$ where f_r is a monotonic transformation estimated from the data. Three types of transformation are allowed:

(a)   *Scale*

$$f_r(\ln d_{ijr}) = \ln d_{ijr} + v_r \qquad (5.26)$$

The constant v_r, estimated separately for each subject, changes the location or origin of the log dissimilarity data. This changes the scale or unit of measurement for the dissimilarity data. Thus each raw dissimilarity quantity is multiplied by the quantity $\exp(v_r)$ and this quantity is termed the *regression* coefficient for the dissimilarities of the rth subject.

(b)   *Power*

$$f_r(\ln d_{ijr}) = p_r \ln d_{ijr} + v_r \qquad (5.27)$$

This estimated transformation amounts to changing both the scale and the location of the log dissimilarities. The effect is equivalent to taking the d_{ijr} to the power p_r and multiplying it by the regression coefficient $\exp(v_r)$.

Ramsay recommends the use of this particular transformation in most applications because it captures at least some of the nonlinearity typically observed between dissimilarity and distance.

(The rationale for using log dissimilarities rather than applying transformations directly to the raw dissimilarities is discussed in Ramsay, 1982.)

Table 5.10 Ratings of dissimilarity for 15 recreations

	1	2	3	4	5	6	7	8	9	10	11	12	13	14	15	16
A	—															
B	16	—														
C	3	18	—													
D	12	12	11	—												
E	15	21	16	2	—											
F	20	10	19	15	12	—										
G	15	12	13	9	19	6	—									
H	21	23	23	19	7	22	20	—								
I	7	10	6	18	19	25	15	25	—							
J	19	22	25	22	14	8	22	23	25	—						
K	9	7	13	15	12	19	20	22	8	25	—					
L	22	16	16	19	13	7	13	15	23	13	25	—				
M	7	3	13	12	21	13	10	22	13	12	7	18	—			
N	21	22	22	12	23	21	18	18	21	22	9	22	12	—		
O	8	8	7	9	21	21	2	22	5	25	9	23	10	8	—	

(A) Concert, (B) museum, (C) theatre, (D) movie, (E) watch TV, (F) conference, (G) reading, (H) watch hockey, (I) ballet, (J) political debate, (K) fashion show, (L) documentary film, (M) exhibition, (N) window shopping, (O) restaurant. Reproduced with permission from Scientific Software Inc.

(4) Coordinates, subject weights etc. are estimated by maximum likelihood assuming some form of distribution for the residuals, ϵ_{ijr}, given by

$$\epsilon_{ijr} = f_r(\ln d_{ijr}) - \ln d_{ijr}^* \tag{5.28}$$

The usual assumption is that

$$\epsilon_{ijr} \text{ is distributed } N(0, \sigma_r^2 \gamma_{ij}) \tag{5.29}$$

where σ_r^2 is known as the *subject specific* variance component and γ_{ij} the *pair specific* component.

The estimation procedure is described in detail in Ramsay (1977); here we continue with a description of an application of the method taken from Ramsay (1983). Ten subjects were asked to judge every one of the 105 possible dissimilarities among a set of 15 forms of recreation. A subject recorded these judgements by choosing among 25 categories presented for each pair as follows:

Museum and hockey

Very similar ———————————————————————————— **Very different**

A typical dissimilarity matrix is shown in Table 5.10.

The results of applying Ramsay's **MULTISCAL** model with a power transformation are shown in Tables 5.11–13. The final two-dimensional configuration is shown in Figure 5.19. A more detailed description of this example is given in Ramsay (1983).

5.11 Summary

The literature of multidimensional scaling is now vast and continues to grow at a considerable rate. It has been possible in this chapter to discuss only a

Table 5.11   Final two-dimensional configuration

	Recreation	Coordinate 1	Coordinate 2
A	Concert	−36·92	49·08
B	Museum	−23·97	9·14
C	Theatre	−15·56	46·58
D	Movie	11·22	−31·24
E	Watch TV	7·98	−49·78
F	Conference	101·43	1·34
G	Reading	55·47	30·87
H	Hockey	−5·01	−85·18
I	Ballet	−24·76	63·17
J	Debate	121·11	11·58
K	Fashion	−59·49	−17·42
L	Documentary	57·59	−5·92
M	Exhibition	−39·97	−9·38
N	Window shop	−66·46	−45·70
O	Restaurant	−82·66	32·88

Table 5.12   Subject weights for recreations example

Subject	Coordinate 1	Coordinate 2
1	1·04	0·95
2	0·80	1·17
3	1·19	0·77
4	1·12	0·86
5	0·73	1·21
6	0·82	1·15
7	0·75	1·20
8	1·41	0·06
9	1·34	0·46
10	0·25	1·39

Table 5.13   Regression coefficients and exponents

Subject	Reg. coeff.	Exponent
1	0·55	1·88
2	2·23	1·43
3	1·71	1·55
4	0·70	1·79
5	2·24	1·40
6	0·44	2·02
7	1·52	1·51
8	0·29	2·19
10	0·92	1·70

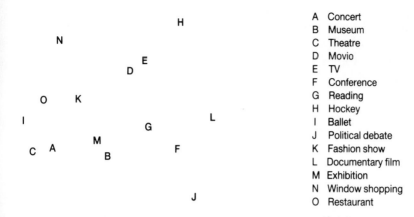

Figure 5.19 MULTISCAL solution on similarities of recreations data

small number of the many methods which have been suggested. Nevertheless it is hoped that the examples presented have demonstrated that the techniques of multidimensional scaling provide powerful possibilities for the analysis of proximity matrices. Many other interesting applications are described in Romney *et al.* (1972), and Davison (1983).

The introduction of stochastic models into the area by Ramsay indicates a maturing of the technology of this type of data analysis, although this has not been universally welcomed as is witnessed by the following quotation, from Chatfield (1982):

> I find myself unconvinced that this sort of model adds very much to our under-standing of this sort of problem and I suggest that this is one area of statistics where the emphasis should remain with the data analytic, exploratory techniques.

Exercises

5.1 Suggest how Kruskal's method of non-metric multidimensional scaling may be adapted for situations (a) when there are missing values in the proximity matrix and (b) when there are ties in this matrix.

5.2 What is meant by the horseshoe effect in multidimensional scaling? (See D.G. Kendall, 1975).

5.3 Apply both metric multidimensional scaling (using a linear transformation of the data) and non-metric multidimensional scaling to the similarity matrix shown in Table 5.14 (taken from Woese, 1981). In each case find a solution in 2, 3 and 4 dimensions. (The results illustrate the problem of degeneracy – see Kruskal and Wish, 1978.)

5.4 Table 5.15 summarizes data collected during a survey during which sub-jects were asked to compare a set of eight legal offences, and to say for each one how unlike it is, in terms of seriousness, compared to each of the others in turn. Each entry in the table shows the percentage of

Table 5.14

·29														
·33	·36													
·05	·10	·06												
·06	·05	·06	·24											
·08	·06	·07	·25	·22										
·09	·10	·07	·28	·22	·34									
·11	·09	·09	·26	·20	·26	·23								
·08	·11	·06	·21	·19	·20	·21	·31							
·11	·10	·11	·06	·11	·12	·11	·14							
·11	·10	·10	·12	·07	·13	·12	·11	·12	·51					
·08	·13	·09	·07	·06	·06	·09	·10	·10	·25	·25				
·08	·07	·07	·12	·09	·12	·10	·10	·12	·30	·24	·32			
·10	·09	·11	·07	·07	·10	·10	·13	·12	·34	·31	·29	·28		
·07	·07	·06	·07	·05	·07	·06	·10	·06	·17	·15	·13	·16	·19	
·08	·09	·07	·09	·07	·09	·09	·10	·07	·10	·20	·21	·23	·23	·13

Table 5.15   (Reprinted by permission of Peters, Fraser and Dunlop Group Ltd.)

Offence	1	2	3	4	5	6	7	8
1	0							
2	21·1	0						
3	71·2	54·1	0					
4	36·4	36·4	36·4	0				
5	52·1	54·1	52·1	0·7	0			
6	89·9	75·2	36·4	54·1	53·0	0		
7	53·0	73·0	75·2	52·1	36·4	88·3	0	
8	90·1	93·2	71·2	63·4	52·1	36·4	73·0	0

(1) Assault and battery, (2) rape, (3) embezzlement (4) perjury, (5) libel, (6) burglary, (7) prostitution, (8) receiving stolen goods

respondents who judge the two offences as very dissimilar. Find a two-dimensional scaling solution and interpret the dimensions underlying the subjects' judgements.

(1) Assault and battery, (2) rape, (3) embezzlement, (4) perjury, (5) libel, (6) burglary, (7) prostitution, (8) receiving stolen goods.

5.5   The four matrices given in Table 5.16 arise from asking four individuals to rate how similar they considered each pair in a set of 10 famous Frenchmen. A rating of '1' indicates that the pair were considered very similar, one of '9' completely dissimilar. Apply some form of multidimensional scaling to the 'averaged' proximity matrix and compare the results with those obtained by using a method of scaling which allows for individual differences.

5.6   What is an asymmetric matrix? Suggest four possible methods for the analysis of asymmetric proximity matrices and, if possible, apply each method to the following data (Table 5.17) which arises from children's ratings of other children in their class. (See Chatfield and Collins, 1980, for details.)

Table 5.16

(a) Subject 1	1	2	3	4	5	6	7	8	9	10
1	1									
2	4	1								
3	7	8	1							
4	8	8	2	1						
5	7	8	8	2	1					
6	1	7	6	8	8	7	1			
7	4	6	6	8	4	5	1			
8	5	8	6	7	3	3	5	1		
9	7	8	2	4	4	8	7	6	1	
10	6	8	3	3	4	8	5	6	2	1

(b) Subject 2	1	2	3	4	5	6	7	8	9	10
1	2									
2	3	6	7							
3	5	6	3							
4	7	9	4	5						
5	6	7	7	3						
6	2	7	6	8	7	1				
7	3	5	8	8	9	6	4			
8	6	5	3	5	9	3	4	2		
9	8	9	3	4	4	7	5	1		
10	6	6	2	2	5	9	6	4	2	

(1) Pompidou, (2) Vignancourt, (3) Ducks, (4) Geismar, (5) Mitterand, (6) De Gaulle (7) Poker, (8) Sebeiber, (9) Seguy, (10) Sartre

Table 5.17

Child	*1*	*2*	*3*	*4*		*6*	*7*	*8*
1	0	25	3	25	1	1	2	3
2	1	0	4	25	2	1	3	4
3	1	25	0	25	1	2	1	2
4	1	25	2	0	2	1	1	2
5	1	25	2	25	0	1	1	2
6	1	25	3	25	2	0	2	3
7	2	25	1	25	1	2	0	1
8	1	25	2	25	2	2	1	0

6

Cluster Analysis

6.1 Introduction

An important component of virtually all scientific research is the classification of the phenomena being studied. In the behavioural sciences, for example, these may be individuals or societies or even patterns of behaviour or perception. The investigator is usually interested in finding a classification in which the items of interest are placed into a small number of homogeneous groups or clusters. These are usually mutually exclusive, but this is not obligatory. At the very least the classification may provide a convenient summary of the multivariate data on which it is based, but often it will yield much more than this. It will be an aid to memory and the understanding of the data, and will facilitate communication between different groups of research workers. Often it will have important theoretical or practical implications. In psychiatry, for example, the classification of mental disturbances should help in the search for their causes and lead to improved methods of therapy.

The simplest approach to discovering distinct groups or clusters is by the examination of scattergrams. These may be obtained by plotting the first two or three principal components (see Chapter 4) or from the results of multidimensional scaling (see Chapter 5). Other exploratory methods useful in the search for clustering are the use of Andrews plots and Chernoff 'faces' (see Chapter 3). Once evidence of clustering has been found it will often be useful to provide some sort of explicit classification using one or more *cluster analysis* algorithms. This is the subject covered in the present chapter. There are now very many different methods of cluster analysis and only a selection of the most widely used methods will be introduced here. Comprehensive reviews of clustering techniques are provided by Cormack (1971) and Everitt (1980).

(An alternative aspect of classification, namely that when the groups are known *a priori* and the aim is to produce a rule for classifying new individuals is taken up in Chapter 12.)

Table 6.1   Some typical questions from the Buss–Durkee questionnaire

I seldom strike back, even if someone hits me first	TRUE/FALSE
I know that people tend to talk about me behind my back	TRUE/FALSE
I demand that people respect my rights	TRUE/FALSE

Table 6.2   Means of the eight variables on each of three clusters derived from the multivariate normal mixture approach

					Variable				
	N	1	2	3	4	5	6	7	8
$c1$	43	5·5	3·5	5·0	2·9	2·9	3·3	6·4	3·9
$c2$	17	5·8	6·8	7·2	2·7	4·8	5·6	9·5	6·2
$c3$	26	6·4	3·1	4·5	3·0	4·6	5·1	7·0	5·3

N = number of observation in a cluster.

6.2   Examples of the Application of Cluster Analysis

Before proceeding to describe any particular method of cluster analysis in detail it may be useful to look at some examples of their application in order to illustrate their utility.

The first example is of an analysis of a set of data consisting of eight scores for each of 86 long-term prisoners in a British prison. The eight scores were derived from the Buss-Durke hostility questionnaire (Buss and Durkee, 1957), which consists of 75 questions having an answer 'true' or 'false.' Some typical questions are shown in Table 6.1. The eight scores used in this investigation are subtotals of different sets of the original seventy-five answers; the scores measure various aspects of hostility, these being as follows:

1. Assault	5. Resentment
2. Indirect Hostility	6. Suspicion
3. Irritability	7. Verbal Hostility
4. Negativism	8. Guilt

The method of cluster analysis to be described in Section 6.4 was applied to these data and a three-cluster solution obtained. Table 6.2 shows the within-cluster means for the eight scores and the number of individuals in each cluster.

The first cluster is characterised by low mean scores on all eight variables, indicating rather low hostility. This could perhaps be termed an essentially 'non-aggressive' group who, on the whole, are likely to be less troublesome for prison authorities. The other two groups show more hostility and could perhaps both be termed 'aggressive' groups. However, they differ in the manner in which their aggression is displayed: Group 2 seem more likely to show verbal hostility rather than to resort to physical assault; in group 3 hostility is likely to manifest itself more violently. Such findings may have important practical implications for the management of prisoners.

A second example concerns a classification of attempted suicide by cluster analysis reported by Paykel and Rassby (1978). The subjects here were 236 suicide attempters presenting at the main emergency service for one city in the USA. From the pool of available variables, fourteen were selected as especially relevant to classification and were used in the cluster analysis. The variables included age, sex, previous suicide history, recent suicidal feelings, and so on. Prior to cluster analysis, a principal components analysis was carried out on the intercorrelation matrix of the fourteen variables, and the first twelve principal component scores used in the cluster analysis rather than the original variables. (This indicates a further possible use of principal components analysis.) Several of the agglomerative hierarchical clustering techniques to be described in the next section were applied to the data and the most satisfactory results were obtained from Ward's method, which indicated that suicide attempters could be classified into three groups. The largest group comprises patients taking overdoses, with less risk to life and a predominance of inter-personal motivations. A second smaller group is distinguished by the use of violent methods with higher risk to life. A third group comprises recurrent attempters, with previous histories of many attempts, relatively low risk to life, and overtly hostile behaviour. Whilst not claiming that these findings provide a definitive classification of suicide attempters, the authors do suggest that it has the virtue of simplicity, and may be worthy of further exploration.

6.3   Agglomerative Hierarchical Clustering Techniques

This family of methods includes some of the oldest and most frequently used clustering techniques. They all operate in essentially the same way, proceeding sequentially from the stage in which each object is considered to be a single member 'cluster,' to the final stage in which there is a single group containing all *n* objects. At each stage in the procedure the number of groups is reduced by one by joining together or *fusing* the two groups considered to be the most similar or the closest to each other. The variety of techniques available arises because of the different possibilities for defining inter-group distance or similarity, as we shall see later.

Since the clusters at any stage are obtained by the fusion of two clusters from the previous stage, these methods lead to a hierarchical structure for the objects. One useful visualisation of such a hierarchy is a *tree diagram*, more commonly known as a *dendrogram*. Examples of such diagrams will be given later. Before discussing any of these techniques in detail, however, we should perhaps consider briefly the implications of imposing a hierarchical structure of this kind on data.

The concept of the hierarchical representation of a data set was developed primarily in biology. The structures output from a hierarchical clustering method resembles the traditional hierarchical structure of Linnaean taxonomy with its graded sequence of ranks. Although any numerical taxonomic exercise with biological data need not replicate the structure of traditional classification, there nevertheless remains a strong tendency among biologists to prefer hierarchical classifications. Hierarchical clustering methods are now used, however, in many other fields in which hierarchical structures may not be the most appropriate, and the logic of their use in such areas needs careful evaluation. For example, in their biological application questions concerning

Table 6.3 Dissimilarity matrix for five individuals

$$
\mathbf{D} = \begin{array}{c} \\ 1 \\ 2 \\ 3 \\ 4 \\ 5 \end{array}
\begin{array}{ccccc}
1 & 2 & 3 & 4 & 5 \\
\left(\begin{array}{ccccc}
0.0 & & & & \\
2.0 & 0.0 & & & \\
6.0 & 5.0 & 0.0 & & \\
10.0 & 9.0 & 4.0 & 0.0 & \\
9.0 & 8.0 & 5.0 & 3.0 & 0.0
\end{array}\right)
\end{array}
$$

the optimal number of groups do not arise — here the investigator is often specifically interested in the complete tree structure. Such questions, are however often raised by other users of these techniques, who consequently require a decision regarding the stage of the hierarchical clustering process at which an optimum partitioning of the items to be classified is obtained. This important and difficult question will be discussed in Section 6.3.3.

6.3.1   Measuring Inter-Cluster Dissimilarity

Agglomerative hierarchical techniques differ primarily in how they measure the distance or similarity of two clusters (where a cluster may, at times, consist of only a single object). Perhaps the simplest inter-group measures are

$$d_{AB} = \min_{\substack{i \in A \\ j \in B}}(d_{ij}) \tag{6.1}$$

$$d_{AB} = \max_{\substack{i \in A \\ j \in B}}(d_{ij}) \tag{6.2}$$

where d_{AB} is the dissimilarity between two clusters A and B, and d_{ij} is the dissimilarity between objects i and j. (This, of course, could be one of a large variety of measures, including, for example, Euclidean and city block distances.) The dissimilarity measure in equation (6.1) is the basis of *single linkage* clustering and that in (6.2) the basis of *complete linkage* clustering. Both these techniques have the often desirable property that they are invariant under monotone transformations of the original inter-object dissimilarities (cf. non-metric multidimensional scaling, Section 5.4). To illustrate the operation of agglomerative hierarchical clustering techniques we shall apply both single and complete linkage to the dissimilarity matrix shown in Table 6.3.

6.3.2   Single Linkage Clustering

At stage one of the procedure, individuals 1 and 2 are merged to form a cluster, since d_{12} is the smallest entry in the dissimilarity matrix $\mathbf{D}$. The distances between this group and the three remaining single individuals 3,4 and 5 are obtained from $\mathbf{D}$ as follows:

$$d_{(12)3} = \min(d_{13}, d_{23}) = d_{23} = 5.0,$$
$$d_{(12)4} = \min(d_{14}, d_{24}) = d_{24} = 9.0,$$
$$d_{(12)5} = \min(d_{15}, d_{25}) = d_{25} = 8.0$$

We may now form a new distance matrix $\mathbf{D}_1$, giving inter-individual dissimilarities, and cluster-individual dissimilarities:

$$
\mathbf{D}_1 = \begin{array}{c} \\ (12) \\ 3 \\ 4 \\ 5 \end{array}
\begin{array}{c} \begin{array}{cccc} (12) & 3 & 4 & 5 \end{array} \\
\left(\begin{array}{cccc}
0.0 & & & \\
5.0 & 0.0 & & \\
9.0 & 4.0 & 0.0 & \\
8.0 & 5.0 & 3.0 & 0.0
\end{array} \right)
\end{array}
$$

The smallest entry in $\mathbf{D}_1$ is d_{45} and so individuals 4 and 5 are now merged to form a second cluster, and the dissimilarities now become

$$d_{(12)3} = 5.0 \text{(as before)},$$
$$d_{(12)(45)} = \min(d_{14}, d_{15}, d_{24}, d_{25}) = d_{25} = 8.0$$
$$d_{(45)3} = \min(d_{34}, d_{35}) = d_{34} = 4.0$$

These may be arranged in a matrix $\mathbf{D}_2$ where

$$
\mathbf{D}_2 = \begin{array}{c} \\ (12) \\ 3 \\ (45) \end{array}
\begin{array}{c} \begin{array}{ccc} (12) & 3 & (45) \end{array} \\
\left(\begin{array}{ccc}
0.0 & & \\
5.0 & 0.0 & \\
8.0 & 4.0 & 0.0
\end{array} \right)
\end{array}
$$

The smallest entry is now $d_{(45)3}$ and so individual 3 is added to the groups containing individuals 4 and 5. Finally, fusion of the two groups at this stage takes place to form a single group containing all five individuals. The dendrogram illustrating this series of mergers is shown in Figure 6.1.

This technique seems first to have been described by Florek *et al.* (1951) under the title 'dendritic method'. McQuitty (1957) and Sneath (1957) independently introduced slightly different versions of it, and it was the subject of further discussion in Johnson (1967).

6.3.3 Complete Linkage Clustering

As with single linkage this method begins by merging individuals 1 and 2. The dissimilarities between this cluster and the three remaining individuals 3, 4 and 5 are obtained from $\mathbf{D}$ as follows:

$$d_{(12)3} = \max(d_{13}, d_{23}) = d_{13} = 6.0,$$
$$d_{(12)4} = \max(d_{14}, d_{24}) = d_{14} = 10.0,$$
$$d_{(12)5} = \max(d_{15}, d_{25}) = d_{15} = 9.0$$

The final result is the dendrogram shown in Figure 6.2.

A further possibility for measuring inter-cluster dissimilarity is the following

$$d_{AB} = \frac{1}{n_A n_B} \sum_{i \in A} \sum_{j \in B} d_{ij} \tag{6.3}$$

where n_A and n_B are the number of individuals in clusters A and B. This measure is the basis of a very widely used procedure known as *group average clustering*. Another very popular technique is that introduced by Ward (1963), who proposed that at any stage of the analysis the loss of information which results from the grouping of objects into clusters can be measured by the total sum of squared deviations of every object's variable values from their

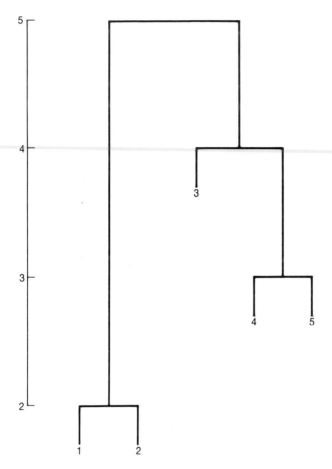

Figure 6.1 Single linkage dendrogram

respective cluster means. At each step in the analysis, union of all possible pairs of clusters is considered, and the two clusters where fusion results in the minimum increase in the sum of squares are combined. A numerical example illustrating group average clustering is given in Everitt (1977b), and one for Ward's technique in Everitt (1980). It is important to note that neither of these techniques is invariant to monotone transformations.

6.3.4　An Example of the Application of Single Linkage, Complete Linkage, and Group Average Clustering

Each of these methods was applied to the similarity matrix of thirty adjectives describing pain given in Table 5.6. The resulting dendrograms appear in Figures 6.3, 6.4 and 6.5.

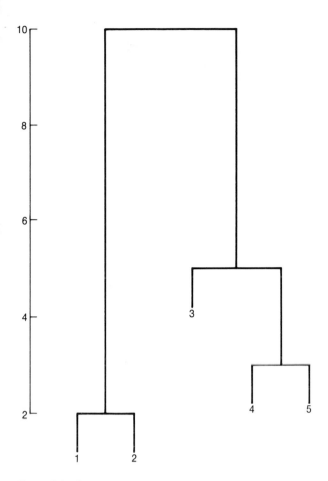

Figure 6.2 Complete linkage dendrogram

For these data the results given by each of the three methods are very similar and the clusters of adjectives appear intuitively sensible, with, for example, words such as flickering, quivering, jumping and flashing forming one group, and hot, burning and scalding forming another. The detailed analyses of these data, and a comparison of the cluster analysis grouping with those proposed originally by Melzack (1975), are given in Reading *et al.* (1982).

These results can usefully be combined with those obtained from a multi-dimensional scaling of the same data (see Section 5.6) by indicating part of the hierarchical cluster structure on the two-dimensional solution obtained from the scaling procedure. Figure 6.6 shows the single linkage solution displayed in this way, and provides a convenient and informative display of the relationships between the adjectives as indicated by their similarities.

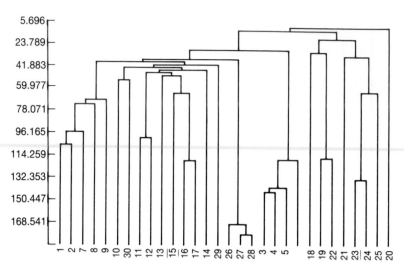

Figure 6.3   Single linkage dendrogram for pain adjectives similarity matrix

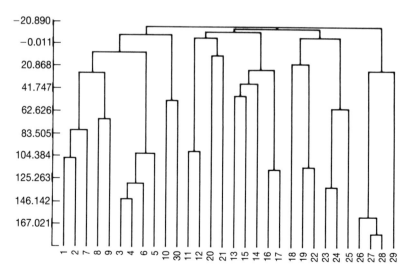

Figure 6.4   Complete linkage dendrogram for pain adjectives similarity matrix

As a further illustration of the application of these methods both complete linkage and average linkage were applied to the athletic records data given in Chapter 4, using Euclidean distances between countries on the times for each event in seconds. The resulting two dendrograms are shown in Figures 6.7 and 6.8; for these data the two approaches give similar results. Relating

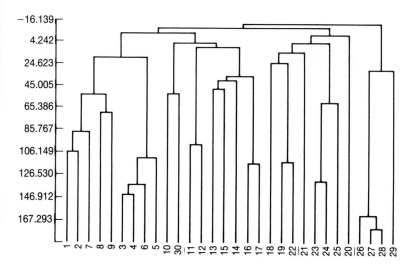

Figure 6.5   Group average dendrogram for pain adjectives similarity matrix

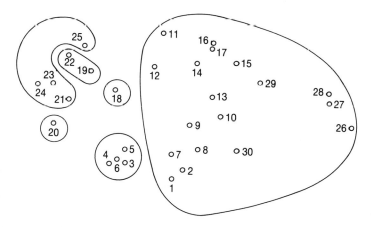

Figure 6.6   Single linkage cluster solution for pain adjectives embedded in two-dimensional non-metric multidimensional scaling solution

the results of the cluster analysis to the principal components plot and faces representation of these data (see Chapter 4) is left as an exercise for the reader.

6.3.5   Measuring Goodness of Fit

Once a dendrogram has been obtained from some particular hierarchical technique it is important to consider how well it 'fits' the original similarity

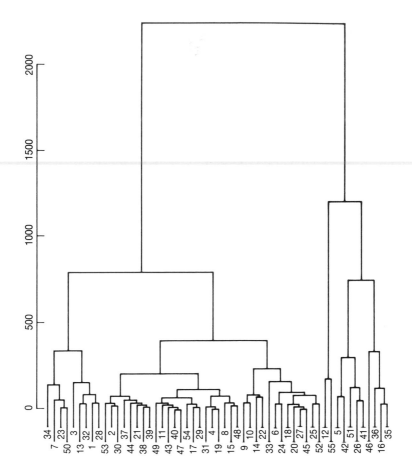

Figure 6.7 Complete linkage dendrogram for athletic records data

or dissimilarity matrix. In this way we might be able to assess whether the derived hierarchical structure is suitable for the data set under investigation. Essentially there are two aspects of fit that require attention. The first is the global fit of the dendrogram for the input similarities or dissimilarities. The second is concerned with which particular partition obtained from the hierarchy is, in some sense, optimal. In other words, what is the best number of groups?

The usual way to assess the global fit of the dendrograms is by use of the *cophenetic correlation coefficient* (CPCC). This is simply the product moment correlation between the entries of the dissimilarity matrix and those of the so-called *cophenetic matrix*, the elements of the latter being the fusion level at which a pair of objects appear together in the same cluster for the first time. For example, the cophenetic matrix arising from the application of single

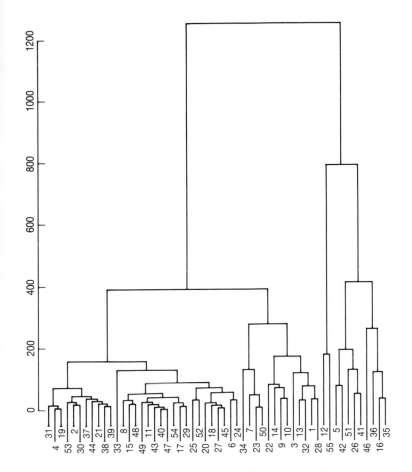

Figure 6.8 Group average dendrogram for athletic records data

linkage to the dissimilarity matrix in Table 6.3 is given in Table 6.4. The cophenetic correlation for this example gives the value 0.82.

In general, 'large' values of the CPCC are taken as indicating that the dendrogram provides a reasonable summary of the observed similarities or dissimilarities. However, the question obviously arises as to what, exactly, constitutes a 'large' value. In an attempt to answer this question Rohlf and Fisher (1968) studied the distribution of the CPCC under the null hypothesis of a single cluster versus the alternative hypothesis of a system of nested clusters. They found that a value of the CPCC above 0.8 is needed for rejecting the null hypothesis, although in a later paper, Rohlf (1970) warns that even a CPCC near 0.9 does not guarantee that the dendrogram provides an adequate summary of the observed dissimilarities. In other words, very high values of the CPCC are needed before it can be safely assumed that a hierarchical cluster structure is suitable for the data under investigation. When one is

Table 6.4　Cophenetic matrix obtained by applying single linkage clustering to Table 6.3

$$
\mathbf{C} = \begin{array}{c} 1 \\ 2 \\ 3 \\ 4 \\ 5 \end{array}
\begin{pmatrix}
0.0 & & & & \\
2.0 & 0.0 & & & \\
5.0 & 5.0 & 0.0 & & \\
5.0 & 5.0 & 4.0 & 0.0 & \\
5.0 & 5.0 & 4.0 & 3.0 & 0.0
\end{pmatrix}
$$

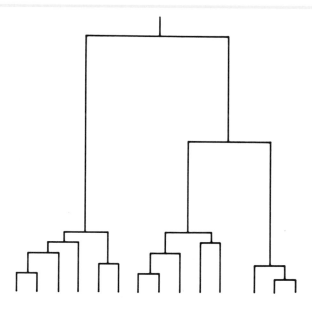

Figure 6.9　Dendrogram suggestive of three groups

only willing to assume that the observed similarities or dissimilarities have ordinal significance, then a coefficient of rank correlation is used to measure the global fit of the dendrogram in place of the CPCC; for example, Hubert (1974) proposed the Goodman–Kruskal γ statistic for this purpose.

With hierarchical clustering, partitions for a particular number of groups are obtained by selecting one of the clusterings in the nested sequence of groupings that comprise the hierarchy. Many investigators will be interested in selecting that partition which best fits the data in some sense and a number of possible methods for this have been suggested. Perhaps the most common is to examine the dendrogram for large changes between adjacent fusion levels; such a change in going from, say j to $j - 1$ groups might be indicative of a j group solution. For example, Figure 6.9 shows a dendrogram rather suggestive of three groups.

Such a procedure may, of course, be rather subjective as some of the examples in Everitt (1980), Chapter 5, indicate. Mojena (1977) suggests a more objective procedure based upon the relative sizes of the different fusion levels. His

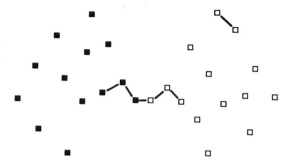

Figure 6.10 Data illustrating chaining effect

proposal is to select the number of groups corresponding to the first stage in the dendrogram satisfying

$$\alpha_{j+1} > \bar{\alpha} + ks_\alpha \tag{6.4}$$

where $\alpha_0, \alpha_1, \ldots, \alpha_{n-1}$ are the fusion levels corresponding to stages with $n, n - 1, \ldots, 1$ clusters, $\bar{\alpha}$ and s_α are, respectively, the mean and unbiased standard deviation of the α values, and k is a constant. As given in (6.4) the stopping rule is for fusion levels arising from the analysis of dissimilarity matrices; in the case of similarities where $\alpha_0 > \alpha_1 \cdots > \alpha_{n-1}$ the corresponding inequality is $\alpha_{j+1} < \bar{\alpha} - ks_\alpha$. Mojena suggested that values of k in the range 2.75 to 3.50 gave the best overall results. If no values of α satisfies inequality (6.4) then the data are regarded as consisting of a single cluster. Mojena gives a number of examples which indicate that his suggested procedure may prove very useful and have many practical applications. (Other rules for selecting a particular number of groups from a dendrogram are discussed by Milligan, 1985.)

6.3.6 Some Properties of Agglomerative Hierarchical Clustering Techniques

Single linkage often does not give satisfactory results if intermediates are present between clusters, due to the phenomenon known as *chaining*, which refers to the tendency of a method to incorporate these intermediate points into an existing cluster rather than initiating a new one. A set of two-dimensional data which might give rise to chaining is shown in Figure 6.10. Because of this problem single linkage tends to lead to the formation of long straggly clusters.

Group average, complete linkage and Ward's method often find spherical clusters even when the data appear to contain clusters of other shapes. Consequently they may impose a structure on the data rather than extract the actual structure present. This problem is illustrated for some two-dimensional data in Figure 6.11.

In the late 1960s there were several attempts at constructing a theoretical framework within which to study the properties of hierarchical clustering techniques. For example, Johnson (1967) showed that the entries in the cophenetic

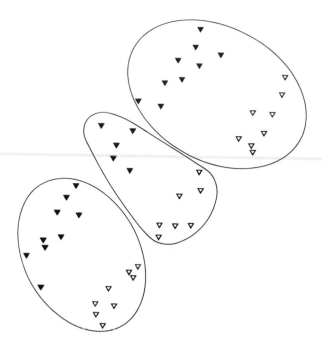

Figure 6.11 Two-dimensional data illustrating type of solution given by group average, Ward's method and complete linkage for elliptical groups

matrices derived from such methods satisfy the *ultrametric inequality,*

$$d(x, y) \leq \max[d(x, z), d(y, z)] \qquad (6.5)$$

(see Exercise 6.1). Johnson showed that each hierarchical clustering scheme gives rise to this particular kind of metric, and, conversely, that given such a metric we may recover the hierarchical structure from it.

Jardine and Sibson (1968) took this as their starting point and argued that since the input proximities are not generally ultrametric (and only occasionally metric), then a cluster method which transforms a proximity matrix into a hierarchic dendrogram should be regarded as a method whereby the ultrametric inequality is imposed on a proximity measure. They then specify a number of criteria which, they argue, should be met by any such transformation. The first three criteria are as follows:

(1)   A unique result should be obtained from a given proximity matrix; that is, the transformation should be *well defined.*

(2)   Small changes in the data should produce small changes in the resultant dendrograms; that is, the transformation should be *continuous.*

(3)   If the dissimilarity coefficient is already ultrametric it should be unchanged by the transformation.

Jardine and Sibson show that these criteria are satisfied only by the single linkage method and, consequently, suggest that other hierarchical techniques are unacceptable. Since single linkage has never proved to be a very popular or successful method in practice, this recommendation has led to some controversy, the most severe criticism coming from the so-called 'Australian school' of Williams, Lance, Dale and Clifford (1971).

Consider first the criterion that the transformation should be well defined; that is a unique result should be obtained from a given proximity matrix. Methods other than single linkage will fail this test only when there are equal proximity values which are arbitrarily resolved. Williams *et al.* suggest that in practice such ambiguities are rare, but where they are encountered propose an alternative to making an arbitrary choice, involving reference to information outside the ambiguity under consideration, in particular to the relationships with all other entities in the proximity matrix.

Williams *et al.* also produce fairly convincing arguments against the need for these techniques to satisfy the criteria of being continuous and leaving an ultrametric unchanged by transformation. Essentially they are more concerned with a pragmatic approach to numerical classification and have obviously found methods other than single linkage useful in practice. Gower (1975) also feels that Jardine and Sibson's rejection of all but single linkage clustering is too extreme and that their criteria may be too stringent. He concludes that some of the criteria are not essential.

It must be said that the approach taken by Jardine and Sibson appears to have had little impact on the majority of users of cluster analysis. Single linkage is not popular with investigators employing clustering methods, and the alternative mathematically acceptable method provided by these two authors is applicable only to small data sets, and the solutions given are generally rather difficult to interpret.

An alternative to setting up a series of mathematical criteria for studying and evaluating the numerous agglomerative hierarchical techniques is to compare their effectiveness across a variety of data sets generated to have a particular structure. In this way the solutions obtained by a particular technique may be compared with the generated structure. A number of studies of this type have been undertaken (for example, Cunningham and Ogilvie, 1972). Whilst the results are dependent on the data generated, group average clustering and Ward's method of clustering appear to perform fairly well overall. (These Monte Carlo studies are reviewed by Milligan, 1981.)

6.4   Mixture Models for Cluster Analysis

Although cluster analysis techniques have generally been considered to be primarily 'exploratory' in nature, various attempts have been made to introduce a more formal probabilistic approach. For example, Ling (1973) and Smith and Dubes (1980) consider various probability models for proximity matrices. Here however, we shall consider an approach to clustering the raw data matrix which uses a particular type of probability density function, a *finite mixture density*, as its model. One great advantage of this approach is that it completely dispenses for the need to decide which inter-individual similarity measure is appropriate.

To introduce this approach consider taking a random sample of people living in London and recording for each member of the sample their height. What might be a sensible model for the distribution of this variable in the population? First we have to allow for our sample containing both males and females, since it is well known that, on average males are taller than females. Within each sex it might be reasonable to assume that height is normally distributed with a particular mean and variance. Such considerations lead naturally to the following density function for height:

$$f(\text{height}) = pN(\mu_f, \sigma_f) + (1-p)N(\mu_m, \sigma_m) \qquad (6.6)$$

where p is the proportion of females in the population. The parameters μ_f, σ_f, μ_m, σ_m represent respectively, the mean and standard deviation of height for females and males. A density function of the form given in (6.6) is an example of a class of density functions known as *finite mixtures*. (See Everitt and Hand, 1981 and Titterington *et al*, 1985). In our particular example the main concern would be to use the sample of recorded heights to estimate the five parameters of the density function, namely p, μ_f, σ_f, μ_m and σ_m. Of course, if we had been sensible enough to record the sex of each member of the sample, estimation of these quantities would have been straightforward; here this could have been done very easily. In other areas, however, sexing of species is more difficult and estimation of the parameters from the unlabelled sample becomes a practical neccessity.

The mixture density given by (6.6) involves two univariate, normal components. It is useful as a model for the existence of *two* groups in *univariate* data where the single variable is continuous. The extension to more than two groups is relatively simple, involving the mixture given by

$$f(x) = \sum_{i=1}^{c} p_i N(\mu_i, \sigma_i) \qquad (6.7)$$

where c is the number of groups assumed. A total of $3c - 1$ parameters now need estimating.

Extending the model to deal with *multivariate data* is also simple in principle; in (6.7) the univariate normal components of the mixture are replaced by the corresponding multivariate densities with mean vectors μ_i and covariance matrices Σ_i

$$f(x) = \sum_{i=i}^{c} p_i MVN(\mu_i, \Sigma_i) \qquad (6.8)$$

Now there are $c - 1$ mixing proportions, cd means and $cd(d+1)/2$ variances and covariances to estimate. Clearly estimation of such a large number of parameters is going to be a formidable computational problem, but it can be handled in most cases by maximum likelihood methods, details of which are given in Everitt and Hand (1981). Numerical examples of this approach to clustering are given below.

If the variables being recorded are binary rather than continuous then a mixture density model based upon normal components would obviously not be realistic. The same general approach can however still be used but with a different choice of component density. One possibility is to assume that, within each cluster, responses to individual binary items are independent, with probabilities which are constant within clusters but different between clusters. If we make such an assumption what is the probability density function of variables within a particular group? To answer this question let us begin with an example in which there are three binary variables, x_1, x_2 and x_3, and that within a particular group j, the probabilities of a positive response for each of the variables are θ_{ji}, θ_{j2} and θ_{j3}. Assuming that the three variables are independent of one another within this group we can now find the probability of observing any value of the vector $\mathbf{x}' = [x_1, x_2, x_3]$. For example,

$$Pr[\mathbf{x}' = (0, 1, 1)] = (1 - \theta_{j1})\theta_{j2}\theta_{j3} \tag{6.9}$$

or

$$Pr[\mathbf{x}' = (1, 0, 0)] = \theta_{j1}(1 - \theta_{j2})(1 - \theta_{j3}) \tag{6.10}$$

Both (6.9) and (6.10) may be rewritten in the form

$$Pr[\mathbf{x}'] = (x_1, x_2, x_3) = \theta_{j1}{}^{x_1}(1 - \theta_{j1})^{1-x_1}\theta_{j2}{}^{x_2}(1 - \theta_{j2})^{1-x_2}\theta_{j3}{}^{x_3}(1 - \theta_{j3})^{1-x_3} \tag{6.11}$$

$$= \prod_{l-1}^{3} \theta_{j1}{}^{x_l}(1 - \theta_{j1})^{1-x_l} \tag{6.12}$$

This is known as a *multivariate Bernoulli density* and can be extended to the situation with p binary variables in an obvious fashion:

$$Pr[\mathbf{x}'] = Pr[(x_1, x_2, \ldots, x_p)] = \prod_{l=1}^{p} \theta_{j1}{}^{x_l}(1 - \theta_{j1})^{1-x_l} \tag{6.13}$$

Such density functions now take the place of the normal components of (6.8). Parameter estimation is again by maximum likelihood; for details see Goodman (1974) and Everitt and Hand (1981). (Finite mixture densities based upon components with the form given in (6.13) are essentially equivalent to the *latent class model* proposed by Lazarsfeld and Henry, 1968.) Numerical examples of this approach to the clustering of binary data are given below.

In the behavioural sciences, data are often a mixture of continuous and categorical variables. A finite mixture model appropriate for such data is described by Everitt (1988) and further elaborated in Everitt and Merette (1990).

6.4.1 Some Numerical Examples of the Application of Mixture Distributions

Our first example concerns the application of the multivariate normal mixture model in (6.8) to the data in Table 6.5, which gives murder/manslaughter and rape rates for 16 cities in USA. Since in the case the data involve only two variables they may be plotted as shown in Figure 6.12 such a diagram will allow the results from the mixture analysis to be compared with those obtained by 'visual' analysis.

We shall begin by fitting a two-component bivariate normal mixture using maximum likelihood methods and the EM algorithm (see Everitt and Hand, 1981, for details). Starting values for the parameters were obtained from a 'k means' clustering procedure (see Everitt, 1980, Chapter 3), and the final parameter values were obtained after 12 iterations of the estimation algorithm. (In

Table 6.5   City crime data

	City	Murder	Rape
1	Atlanta	16·5	24·8
2	Boston	4·2	13·3
3	Chicago	11·6	24·7
4	Dallas	18·9	34·2
5	Denver	6·9	41·5
6	Detroit	13·0	35·7
7	Hartford	2·5	8·8
8	Honolulu	3·6	12·7
9	Houston	16·8	26·6
10	Kansas City	10·8	43·2
11	Los Angeles	9·7	51·8
12	New Orleans	10·3	39·7
13	New York	9·4	19·4
14	Portland	5·9	23·0
15	Tucson	5·1	22·9
16	Washington	12·5	27·6

Source United States Statistical Abstract (1970) ; per 100 000 population

this example we have assumed that the correlation between the two variables, murder rate and rape rate is the same in each group.) The results are shown in Table 6.6. The final parameter estimates shown in this table may now be used to find estimates of the posterior probabilities of each city belonging to each of the component densities in the mixture (see Wolfe, 1970). These are given in Table 6.7. The maximum posterior probabilities can be used to partition the cities into two groups. This partition is shown in Figure 6.13. Here the two groups differ predominantly in the murder/manslaughter rate with those in the first group having very high values.

The multivariate normal mixture model was also applied to a set of data described by Powell, Clark and Bailey (1979). These data consist of four test scores obtained from 86 aphasic cases. The four variables were measures of auditory disturbance, usual and reading disturbances, speech and language disturbances and visuomotor and writing disturbances. Fitting the model in (6.8) assuming equal covariance matrices in each group resulted in a four group solution being selected by the likelihood ratio test (see Section 6.4.2). Figure 6.14 shows the means of each group for each of the four variables. Essentially the four groups appear to differ only in severity, and could be contrasted with the original classification of aphasia described by Schuell (1965) in which it was claimed there were seven groups differing in their pattern of symptoms.

The next example relates to fitting the mixture density based on components as given in (6.13) to sets of binary data collected on psychiatric patients during an investigation of social networks. Each patient was asked to give the names of all their acquaintances and for each name supplied to say whether they regarded the person as a friend or not, somebody they could confide in or not, somebody they would miss or not and finally whether the acquaintance was an active one or not. Table 6.8 gives the counts of the number of names falling into each of the 16 possible categories. Latent class analysis was applied and two-class and three-class models fitted. The results are shown in Table 6.9. In

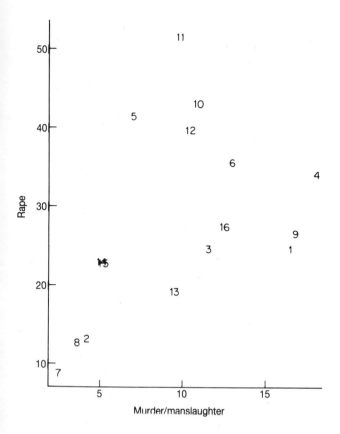

Figure 6.12 Plot of murder/manslaughter and rape rates for 16 cities in the USA

Table 6.6 Results of fitting a two-component bivariate normal mixture to the city crime data using maximum likelihood

Starting values	Final values
$\hat{p} = 0\cdot438$	$\hat{p} = 0\cdot385$
$\hat{\mu}_1' = [13\cdot99, 27\cdot57]$	$\hat{\mu}_1' = [14\cdot40, 27\cdot97]$
$\hat{\mu}_2' = [6\cdot46, 28\cdot54]$	$\hat{\mu}_2' = [6\cdot83, 28\cdot21]$
$\hat{\sigma}_1 = 2\cdot94, \hat{\sigma}_2 = 11\cdot67$	$\hat{\sigma}_1 = 3\cdot00, \hat{\sigma}_2 = 11\cdot68$
$\hat{\rho} = 0\cdot76$	$\hat{\rho} = 0\cdot70$

the former the division is clearly into 'close friends' and perhaps 'just names'. In the three-group solution this division is again clear, with the additional group being 'friends in whom one would not confide'. This description of the

Table 6.7 Estimated posterior probabilities of each city in Table 6.5 belonging to each of the components of the mixture density fitted

		Pr(component 1)	Pr(component 2)
1	Atlanta	1·00	0·00
2	Boston	0·00	1·00
3	Chicago	0·91	0·09
4	Dallas	1·00	0·00
5	Denver	0·00	1·00
6	Detroit	0·78	0·22
7	Hartford	0·00	1·00
8	Honolulu	0·00	1·00
9	Houston	1·00	0·00
10	Kansas City	0·00	1·00
11	Los Angeles	0·00	1·00
12	New Orleans	0·01	0·99
13	New York	0·54	0·45
14	Portland	0·00	1·00
15	Tucson	0·00	1·00
16	Washington	0·95	0·05

On the basis of these probabilities the two groups consist of the following cities:
Group 1. 1, 3, 4, 6, 9, 13, 16
Group 2. 2, 5, 7, 8, 10, 11, 12, 14, 15

data proved extremely useful in other investigations and more details are given in Dunn *et al.* (1989).

6.4.2   Testing for the Number of Components in Mixture Models

Maximum likelihood estimation for both mixtures of multivariate normal densities and for latent class analysis assume that c, the number of components in the mixture (or the number of clusters in the population) is known *a priori*. In many applications however, this will not be so and, consequently, we would need to construct some kind of hypothesis testing procedure for c. One test, considered by a number of authors, including Wolfe (1971) and Binder (1978) is a likelihood ratio test of $c = c_0$ against $c = c_1$. The test statistic λ is given by

$$\lambda = 2(L_{c_1} - L_{c_0}) \tag{6.14}$$

where L_{c_0} and L_{c_1} are the log-likelihoods under the hypotheses of c_0 and c_1 groups. Under the null hypothesis $c = c_0$ λ is generally assumed to be asymptotically distributed as chi-square with degree of freedom equal to the difference in the number of parameters between the two hypotheses. In the context of finite mixtures, however, various authors, for example Everitt (1981, 1988) and McLachlan and Basford (1988) , have shown that λ does *not* have such a null distribution and have suggested various small amendments to the test. Other possibilities which have potential in this area are the use of Akaike and Schwartz's information criteria; see Moore (1989).

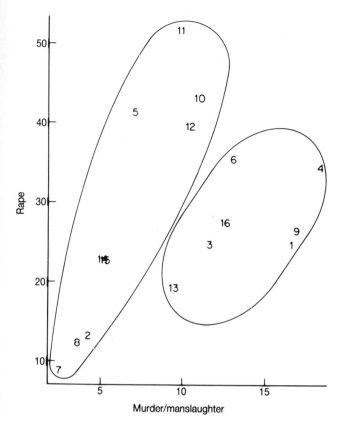

Figure 6.13 Two groups of cities given by estimated posterior probabilities found after fitting two-component bivariate normal mixture to city crime data

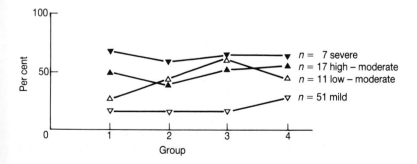

Figure 6.14 Mean error scores (as % of total possible errors) for four-cluster solution on aphasic patients. (Taken with permission from Powell *et al.*, 1979)

Table 6.8 Data collected from long-stay
psychiatric patients

(1)	(2)	(3)	(4)	Frequency
1	1	1	1	529
1	1	1	2	424
1	1	2	1	51
1	1	2	2	193
1	2	1	1	185
1	2	1	2	274
1	2	2	1	46
1	2	2	2	311
2	1	1	1	81
2	1	1	2	279
2	1	2	1	13
2	1	2	2	228
2	1	1	1	25
2	2	1	2	256
2	2	2	1	18
2	2	2	2	1893

(1) active/not active
(2) confides/would not confide
(3) friend/not friend
(4) missed/not missed
1 = first category
2 = second category

Table 6.9 Latent class results for psychiatric data in Table 6.8

(a) Two-class solution

	p	Pr(Active)	Pr(Confides)	Pr(Friend)	Pr(Missed)
Class 1	0·56	0·13	0·00	0·11	0·10
Class 2	0·44	0·78	0·44	0·71	0·84

(b) Three-class solution

	p	Pr(Active)	Pr(Confides)	Pr(Friend)	Pr(Missed)
Class 1	0·41	0·08	0·01	0·02	0·01
Class 2	0·25	0·89	0·73	0·73	0·90
Class 3	0·34	0·48	0·03	0·53	0·58

6.5   Graphical Aids to Cluster Analysis

It has been stressed in the introduction to this text that, in the exploratory
stage of data analysis, graphs and diagrams have an important role to play.
This is particularly so in applications where graphs and diagrams of various
kinds can facilitate the understanding and interpretation of the results obtained
from clustering algorithms.

Some of these aids have already been discussed (for example, scattergrams,
Andrews plots and principal component plots). A number of others will be
described in this section.

6.5.1   Some Sample Plots for Examining Cluster Solutions

Cohen et al. (1977) describes a number of potentially useful *ad hoc* methods for evaluating cluster analysis solutions, some involving very simple plotting procedures. For example, they suggest that a plot of squared distances from certain cluster centroids to entities that are near that centroid might be useful for examining the internal cohesiveness of a cluster. For each cluster centroid, the distances of every entity from that centroid are plotted above the cluster identification shown on the *x*-axis. The symbol plotted is the cluster to which the entity was assigned. Such a plot is shown in Figure 6.15. The information obtainable from this very simple plot includes the following: clusters F and H are extremely well separated from neighbouring entities; an entity not assigned to cluster E is still reasonably close to cluster E, and there is a large distance between the members of cluster A furthest from their centroid and the next closest entities not assigned to A. Another simple graphical aid described by these authors can be used for examining the clusters in terms of either variables used to form the clusters or other variables of interest. Here the clusters are again identified along the *x*-axis, and above each label the values on a certain variable for each entity in that cluster are plotted. The median for the cluster is plotted as a star. This enables one to use the variable in question to compare several entities grouped in the same cluster and to make multiple comparisons across clusters. An example of such a plot appears in Figure 6.16. This shows that clusters A and B are quite similar on this variable except for one entity, A1, in cluster A; cluster E tends to have large entities with moderate spread while cluster M has much smaller values with small spread.

6.5.2   Canonical Variate Plots

A useful method of displaying a set of clusters obtained from the application of some clustering technique is by means of a *canonical variate* plot. Essentially this involves a principal components type analysis of the matrix **G** given by

$$\mathbf{G} = \mathbf{W}^{-1}\mathbf{B} \qquad (6.15)$$

where **W** is the $(p \times p)$ pooled within-groups covariance matrix and **B** is the $(p \times p)$ between-groups covariance matrix. Such an analysis leads to a new set of variables (the canonical variates) which are linear transformations of the original variates that have the properties of being orthogonal and maximising the between-groups variation relative to that within groups. The number of canonical variates that can be derived is equal to the smaller of $c - 1$ and p where c is the number of clusters.

Plots of the data in the space of the first few canonical variates allow a visual inspection of the separation between groups, relative to the variation within groups. For example, Figure 6.17 shows a plot in the space of the two canonical variates of the three groups obtained by applying normal mixture cluster analysis (see Section 6.4) to the 'prisoners' data described in Section 6.2. The major impression gained from this plot is of a 'cloud' of points, with little evidence of any clear-cut 'gaps' between the clusters. The diagram suggests perhaps that this cluster solution has not identified really distinct groups of prisoners, but has merely 'dissected' a homogeneous data set into three parts.

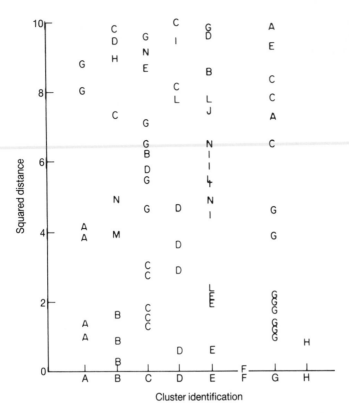

Figure 6.15   Plot of squared distances of selected individuals from their cluster centroids. (Taken with permission from Cohen *et al.*, 1977)

When there are just two groups only one canonical variate is possible, and this is then equivalent to the linear discriminant function, originally derived by Fisher. This function is often used as an aid to classification and diagnosis. For example, suppose a disease can only be diagnosed without error by means of a post-mortem examination. If a number of patients are measured on variables considered to be indicative of the disease, and these patients are then subjected to post-mortem examination upon death, then Fisher's discriminant function may be computed for the two groups — 'with disease' and 'without disease.' Now, given a new patient measured on the same variables, a discriminant function score may be calculated; this may then be compared with a threshold value to judge whether the patient's scores are indicative that the disease is present or not. If the distribution of the scores in each group is multivariate normal with the same variance–covariance matrix, then it can be shown that Fisher's discriminant function is optimal in the sense of minimising the misclassification rate. A detailed account of the method is given in Chapter 12.

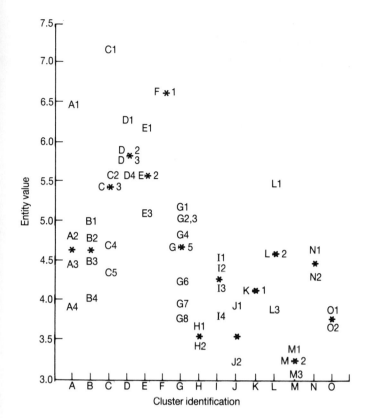

Figure 6.16 Plot of values of a single variable for selected individuals in various clusters (Taken with permission from Cohen *et al.*, 1977)

6.6 Summary

Cluster analysis techniques are potentially very useful for the exploration of complex multivariate data. Its use, however, requires considerable care if misleading solutions are to be avoided, and much attention needs to be given to the evaluation and validation of results. Mixture models present a more 'statistical' approach to the area and in many respects this is to be welcomed; more details of finite mixture distributions are given in Everitt and Hand (1981). Since clustering techniques will generate a set of clusters even when applied to random, unclustered data, the question of validating and evaluating solutions becomes of great importance. A number of generally *ad hoc* procedures for this purpose have been suggested, details of which may be found in Anderberg (1973) and Dubes and Jain (1979).

Finally, we should perhaps comment briefly on the differences between the tree representations of dissimilarity and similarity matrices given by the hierarchical clustering techniques described in this chapter, and the spatial representations given by the methods of multidimensional scaling. Sattath and

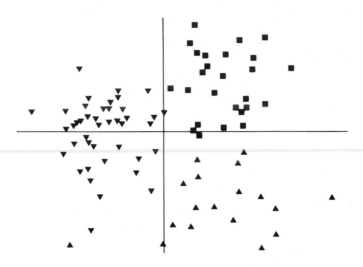

Figure 6.17   Canonical variate plot for 'prisoners' data

Tversky (1977) give some examples where such data are better described by a tree representation rather than by a spatial configuration, but also concede that for some data the opposite will be true. In general, the appropriateness of a tree or a spatial representation will depend upon the nature of both the task and the structure of the stimuli. Some object sets suggest a natural dimensional structure, for example, emotions in terms of intensity and pleasure, and sound in terms of intensity and pitch. Other object sets may suggest a hierarchical structure that results, for example, from an evolutionary process in which all the objects have an initial common structure and later develop additional distinctive features. Perhaps the two approaches are, in general, appropriate for different data, or for capturing different aspects of the same data.

Exercises

6.1   Show that the entries of the cophenetic matrix satisfy the ultrametric inequality.
6.2   Show that the inter-cluster distances used by single linkage, complete linkage and group average clustering satisfy the formula

$$d_{k(ij)} = \alpha_i d_{ki} + \alpha_i d_{kj} + \gamma |d_{ki} - d_{kj}|$$

with

$$\alpha_i = \alpha_j; \quad \gamma = -\frac{1}{2} \quad \text{(single linkage)}$$

$$\alpha_i = \alpha_j; \quad \gamma = \frac{1}{2} \quad \text{(complete linkage)}$$

$$\alpha_i = \frac{n_i}{n_i + n_j}, \quad \alpha_j = \frac{n_j}{n_i + n_j}; \quad \gamma = 0; \quad \text{(group average)}$$

Table 6.10

	Country	Age 0	25	50	75
1	Algeria	53	51	30	13
2	Costa Rica	65	48	26	9
3	El Salvador	56	44	25	10
4	Greenland	60	44	22	6
5	Grenada	61	45	22	8
6	Honduras	59	42	22	6
7	Mexico	59	44	24	8
8	Nicaragua	65	48	28	14
9	Panama	65	48	26	9
10	Trinidad	64	43	21	6
11	Chile	59	7	43	10
12	Ecuador	57	46	25	9
13	Argentina	65	46	24	9
14	Tunisia	56	46	24	11
15	Domin. Rep	64	50	28	11

($d_{k(ij)}$ is the distance between a group k and a group (ij) formed by the fusion of groups i and j, and d_{ij} is the distance between groups i and j.)

6.3 Consider Ward's hierarchical clustering procedure in which clusters are merged so as to produce the smallest increase in the sum-of-squared error terms at each step. Given the ith cluster contains n_i objects and has mean vector $\mathbf{m}_i$ and the jth cluster contains n_j objects with sample mean vector $\mathbf{m}_j$, show that the smallest increase in the error sum-of-squares results from merging the two clusters for which

$$\frac{n_i n_j}{n_i + n_j} d_{ij}^2$$

is a minimum where d_{ij} is the Euclidean distance between the mean of cluster i and j.

6.4 The data in Table 6.10 give expectations of life by country and age for males in 15 countries. Find the estimates of the parameters in a two-component multivariate normal mixture model.

6.5 Respondents in a survey were asked whether or not suicide was acceptable in four different situation, these being as follows:

(1) person has an incurable disease;

(2) person is tired of living;

(3) person has been dishonoured;

(4) person has gone bankrupt.

Responses were coded 1 for yes and 2 for no. The results are summarised in Table 6.11. Fit a latent class model with two classes.

Table 6.11

Response pattern				Frequency
1	2	3	4	
1	1	1	1	105
2	1	1	1	0
1	2	1	1	1
2	2	1	1	0
1	1	2	1	4
2	1	2	1	0
1	2	2	1	4
2	2	2	1	3
1	1	1	2	10
2	1	1	2	1
1	2	1	2	3
2	2	1	2	3
1	1	2	2	62
2	1	2	2	16
1	2	2	2	444
2	2	2	2	724

PART III
REGRESSION MODELS

In this part of the book a distinction is often made between response and explanatory variables, and several alternative types of model are introduced with which to examine the effects of the latter type of variable on the former. The reader should bear in mind that the distinction between the two types of variable is not always clear cut, and that one particular variable might be considered as an explanatory variable for one analysis yet be the response variable in another. The case where one is interested in modelling the variation of a quantitative response variable in terms of variation in one or more quantitative explanatory variables will be familiar to most readers as *multiple linear regression* or, perhaps, *non-linear regression*. Readers should also be familiar with models for the *analysis of variance* (ANOVA) and the *analysis of covariance* (ANCOVA). The *generalized linear model* incorporates all of these linear models and many non-linear ones (see below) as well as *log-linear models* for contingency, tables, *linear-logistic* models for binary response variables, and several *hazard* models for response, failure or survival times.

7

The Generalised Linear Model

7.1 Linear Models

Consider a situation where we are interested, for example, in describing the number of times people visit their doctor annually as a function of their age. There is of course, considerable variation in the number of times people of a given age visit their doctor and so only population means or *expected values* for the number of visits will be considered in this section. Suppose that the average number of visits is predicted by the following equation:

$$\text{Visits} = 2 + \frac{1}{20}\text{Age} \qquad (7.1)$$

This means that, on average, a person aged 40 will visit a doctor four times a year, a person aged 80 six times, and so on. A more general (algebraic) form of this equation is:

$$y = \beta_0 + \beta_1 x \qquad (7.2)$$

Where y is the *response variable* (here, Visits) and x the *explanatory* variable (here, Age). The terms β_0 and β_1 are called *parameters* of the model. The latter have fixed but usually unknown values which have to be estimated from sample data. The parameters of a linear model such as (7.2) are often called *regression coefficients*; they are merely numbers that describe the way in which the response variable is dependent on the explanatory variable. Graphically equation (7.2) can be represented as a single straight line (Figure 7.1). It can be seen that the parameter is the value of y where the line intercepts the vertical axis (that is, when $x = 0$), and β_1 is the slope of the line.

It may now be asked whether, after allowing for the effect of age, a person's sex has any influence on the frequency of visits to a physician. On the assumption that it has the model appropriate to this situation might be the following:

$$\text{Visits} = \beta_0 + \beta_1 \text{Age} + \beta_2 \text{Sex} \qquad (7.3)$$

or, in general,

$$y = \beta_0 + \beta_1 x_1 + \beta_2 x_2 \qquad (7.4)$$

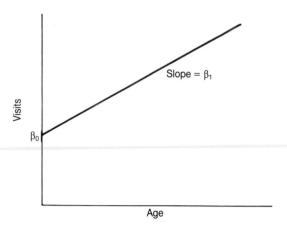

Figure 7.1

This description is not quite so straightforward as before because the variable Sex cannot have the same quantitative meaning as measurement of a person's age. It is here defined to be a *dummy variable* that takes a value of 0 for men, and 1 for women: it is merely a way of distinguishing between males and females. The model in equation (7.3) describes the two lines shown in Figure 7.2. These lines are parallel, with their common slope equal to β_1. The parameter β_0 is the intercept of the male line on the vertical axis, and β_2 is the vertical distance between the two lines. The parameter β_0 (which could of course, be negative) is a measure of the influence of a person's sex on the expected annual number of visits to a doctor. Here, if β_2 is positive then women visit their doctor more often than men, if β_2 is negative then men visit their doctor more often than women. It should be clear to the reader that if the dummy variable Sex were coded as 0 for women and 1 for men then this would simply reverse the sign of the parameter β_2 and change the meaning of β_0. The parameter β_0 would now be the intercept for the female line. The fact that the two lines described by (7.3) or (7.4) are parallel means that the effect of age on frequency of visits is the same for both sexes, or, equivalently, that the known effect of a person's sex is the same at all ages. These are merely different ways of saying that there is no *interaction* between the effects of age and sex.

Now consider a situation where there is interaction between the effects of sex and age. How could this be incorporated into our model? An interaction between the effects indicates that the two lines describing the effect of age on doctor visits for the two sexes are not parallel. A convenient way of expressing this fact algebraically is to create a new variable (Int, or x_3) which is defined as the *product* of the variables Age and Sex (or, in general, of x_1 and x_2). The model describing this situation now becomes

$$\text{Visits} = \beta_0 + \beta_1\text{Age} + \beta_2\text{Sex} + \beta_3\text{Int} \qquad (7.5)$$

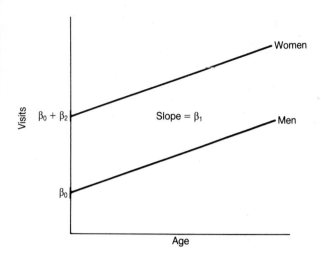

Figure 7.2

or, in general,

$$y = \beta_0 + \beta_1 x_1 + \beta_2 x_2 + \beta_3 x_3 \tag{7.6}$$

Each time a new variable has been introduced into the model an additional parameter has been added. To understand what equation (7.6) means, we just consider the number of visits by men. Here both Sex and Int are equal to 0 and equation (7.6) reduces to

$$\text{Visits} = \beta_0 + \beta_1 \text{Age} \tag{7.7}$$

For women, however, Sex=1 and Int therefore takes the same value as age, so that (7.6) now becomes

$$\text{Visits} = \beta_0 + \beta_1 \text{Age} + \beta_2 + \beta_3 \text{Age} \tag{7.8}$$

or

$$\text{Visits} = (\beta_0 + \beta_2) + (\beta_1 + \beta_3) \text{Age} \tag{7.9}$$

Clearly the parameter β_3 is a measure of the difference between the slopes of the two lines (Figure 7.3). Note that the interpretation of β_2 in (7.6) is different to the interpretation of β_2 given in (7.3). In equation (7.6) β_2 is the difference between the *intercepts* for the two lines (that is, the expected sex difference at Age =0).

The above process is an example of *model building*, and provides an approach by which we find a mathematical or geometric description of the structure in the values of the response variable. (The questions of how well the model describes the data or of estimating the values of the parameters are taken up in later chapters). All the models discussed above involve a *linear combination* of the *parameters* $\beta_0, \beta_1, \beta_2, \ldots$, and consequently are known as *linear models*. It is important that the reader understands this point. They are *not* defined as linear models because they involve linear combinations of the exploratory

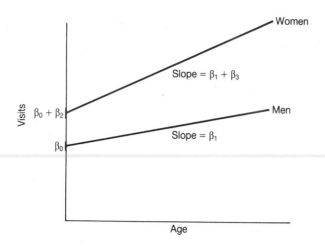

Figure 7.3

variables. Consider the following *polynomial* response model:

$$y = \beta_0 + \beta_1 x + \beta_2 x^2 + \beta_3 x^3 + \cdots \tag{7.10}$$

This is another example of a linear model despite the fact that y is described by a non-linear function of the exploratory variable x. One further example of a linear model involving two quantitative explanatory variables, x_1 and x_2 is given by

$$y = \beta_0 + \beta_1 x_1 + \beta_2 x_2 + \beta_3 x_1^2 + \beta_4 x_2^2 + \beta_5 x_1 x_2 \tag{7.11}$$

Models of the latter type are referred to as quadratic *response surface models*.

Returning to our expected-number-of-visits example let us suppose that we wish to consider now the effect of a person's social class (with, for example, five classes coded as 1, 2, 3, 4 and 5) on our response variable. A possible linear model would be

$$\text{Visits} = \mu + \alpha_i \tag{7.12}$$

where μ is an overall mean for the number of visits, and α_i represents the effect of the ith social class on the number of visits. Equation (7.12) should be familiar to readers as the usual *one-way analysis of variance* model. As written, it appears to be somewhat different from models such as (7.2) and (7.3) but, by extending the concept of dummy variables met earlier, it may easily be rearranged into an essentially similar form. To represent the variable 'social class', five dummy variables are introduced and coded as indicated in Table 7.1. In terms of the five dummy variables, equation (7.12) may be rewritten as

$$\text{Visits} = \mu + \alpha_1 x_1 + \alpha_2 x_2 + \alpha_3 x_3 + \alpha_4 x_4 + \alpha_5 x_5 \tag{7.13}$$

Equation (7.13) is now of the same general form as (7.2), (7.3) and others met earlier in the discussion. As we shall see in more detail in Chapter 8, other analysis of variance models may also be rearranged in this manner.

Table 7.1 Coding of social class in terms of binary dummy variables

Social class	x_1	x_2	x_3	x_4	x_5
		Dummy variables			
1	1	0	0	0	0
2	0	1	0	0	0
3	0	0	1	0	0
4	0	0	0	1	0
5	0	0	0	0	1

The general form of the models discussed in this section is now seen to be

$$y = \beta_0 + \sum_{j=1}^{p} \beta_j x_j \tag{7.14}$$

where y is the expected value of a response variable, and $x_1, x_2, \ldots, x_p$ are p explanatory variables. By introducing a vector $\boldsymbol{\beta}' = [\beta_0, \beta_1, \ldots, \beta_p]$ and another vector $\mathbf{x}' = [1, x_1, x_2, \ldots, x_p]$ Equation (7.14) may be rewritten as

$$y = \mathbf{x}' \boldsymbol{\beta} \tag{7.15}$$

7.2 Non-Linear Models

Linear models are postulated far more often than non-linear ones because they are mathematically easier to manipulate and usually easier to interpret. Fortunately, they appear to provide an adequate description of many data sets. However, there are many other types of mathematical model that could be postulated. An obvious example is to replace a model in which terms are added together by one in which they are multiplied. For the case of a response y being modelled in terms of two explanatory variables, x_1 and x_2 a suitable model might take the form

$$y = \beta_0 x_1^{\beta_1} x_2^{\beta_2} \tag{7.16}$$

Another example of a non-linear model is

$$y = e^{\beta_0 + \beta_1 x_1 + \beta_2 x_2} \tag{7.17}$$

Although it is not a typical property of non-linear models, one interesting characteristic of the models described by equation (7.16) and (7.17) is that they can be converted into a linear form by a suitable mathematical transformation. If we take natural logarithms of both sides of these two equations, we obtain

$$\ln(y) = \ln(\beta_0) + \beta_1 \ln(x_1) + \beta_2 \ln(x_2) \tag{7.18}$$

and

$$\ln(y) = \beta_0 + \beta_1 x_1 + \beta_2 x_2 \tag{7.19}$$

respectively. An example of a non-linear model that cannot be transformed in this way is

$$y = \beta_0 + \beta_1 e^{\beta_2 x_1} + \beta_3 e^{\beta_4 x_2} \tag{7.20}$$

A class of linear models which are of particular practical interest and are derived from a logarithmic transformation of an essentially multiplicative model are those used to describe data arising in the form of a *contingency table*, where a sample of individuals is cross-classified with respect to two or more quantitative variables. The observations are the counts for each cell of the table. A two-dimensional contingency table appears in Table 7.2. For such a table the model of most interest is that of the independence of two classifying variables, which is generally formulated as

$$p_{ij} = p_{i.}p_{.j} \tag{7.21}$$

where p_{ij} is the probability of an observation falling in the ijth cell of the table, and $p_{i.}$ and $p_{.j}$ are row and column marginal probabilities, respectively. (This is the hypothesis usually tested by means of the familiar chi-square test of independence — see, for example, Everitt, 1977a, Chapter 3.) Equation (7.21) may be rewritten in terms of expected cell frequencies, m_{ij} as

$$m_{ij} = N p_{i.}p_{.j} \tag{7.22}$$

where N is the sample size. By taking logarithms, this may be written as

$$\ln(m_{ij}) = \ln(N) + \ln(p_{i.}) + \ln(p_{.j}) \tag{7.23}$$

From here it is a relatively simple matter to show that $\ln(m_{ij})$ may be expressed in the form

$$\ln(m_{ij}) = \beta_0 + \text{Row}(i) + \text{Col}(j) \tag{7.24}$$

where Row (i) and Col (j) are parameters representing the effects of row and column categories, respectively. (For details of how to express equation (7.23) in the form of (7.24), see Everitt, 1977a, Chapter 5, see also Exercise 7.1.) The model in (7.24) is now very similar to that in (7.12) and by a comparable use of dummy variables could be expressed in the same general form as (7.14). Such models are generally known as *log-linear models*, and will be the subject of detailed discussion in Chapter 9. If a typical expected count in a multi-dimensional contingency table is represented by m, then the equivalent of equation (7.14) in this case is

$$m = \exp\left(\beta_0 + \sum_{j=1}^{p} \beta_j x_j\right) \tag{7.25}$$

or

$$\ln(m) = \beta_0 + \sum_{j=i}^{p} \beta_j x_j \tag{7.26}$$

where the model terms are defined as for (7.14).

In many contingency tables there is one variable of particular interest and it is the effects of the remaining variables on this that are of primary interest. (Some examples of such contingency tables appear in Chapter 9). If the former variable is dichotomous, then we might be interested in modelling the probability of an observation falling into one of its categories, p or the ratio of the probabilities for the two categories, $p/(1 - p)$. These quantities might then be modelled as linear functions of the effects of other variables. A typical example might involve a model to predict the proportion of people who are

Table 7.2   A two-dimensional contingency table involving severity of depression and suicidal intent in a sample of 500 psychiatric patients

	Not depressed	Moderately depressed	Severely depressed	Totals
Attempted suicide	26	39	39	104
Contemplated or threatened suicide	20	27	27	74
Neither	195	93	34	322
	241	159	100	500

depressed as a function of their age and sex. For example,

$$p = \beta_0 + \beta_1 \text{Sex} + \beta_2 \text{Age} \tag{7.27}$$

where p is the probability of being depressed, Sex is a dummy variable indicating the person's sex and Age is the person's age. The parameters of the model are β_0, β_1 and β_2. Models such as (7.27) however, have a number of difficulties, the foremost being that the estimated parameter values may lead to fitted p values outside the required range $(0,1)$. (Other problems are described by Cox and Snell, 1989.) An alternative is to model the odds of being depressed, that is $p/(1-p)$ and, in fact, to postulate that the logarithm of this ratio is a linear combination of Sex and Age. For example,

$$\ln(\frac{p}{1-p}) = \beta_0 + \beta_1 \text{Sex} + \beta_2 \text{Age} \tag{7.28}$$

The term on the left-hand of this equation is often referred to as a *logistic transformation* of the proportion p and models such as (7.28) are called *linear-logistic models*. In terms of p itself, the non-linear equivalent to (7.28) is

$$p = \frac{\exp(\beta_0 + \beta_1 \text{Sex} + \beta_2 \text{Age})}{1 + \exp(\beta_0 + \beta_1 \text{Sex} + \beta_2 \text{Age})} \tag{7.29}$$

and in general

$$p = \frac{\exp(\sum_{j=0}^{k} \beta_j x_j)}{1 + \exp(\sum_{j=0}^{k} \beta_j x_j)} \tag{7.30}$$

7.3   Link Functions and Error Distribution in the Generalized Linear Model

In equations (7.1) and (7.2) we describe variation in the expected values of a response variable as a function of a single explanatory variable. The relationship between the quantities has an exact functional form; there is a one-to-one mapping from the person's age to the expected number of visits. If we were now, however, to consider a single individual, Equation (7.2) would no longer be expected to hold, since we know that all not men aged, say 57, will visit a physician the same number of times in a year, simply because of *random variation* in the population of patients. The model applicable to the number of visits for individuals must take some account of this variation. This is usually dealt with by introducing a random disturbance term into the model to give, for example,

$$y_i = \beta_0 + \beta_1 x_i + \epsilon_i \tag{7.31}$$

where y is now the number of visits for the ith member of the population and ϵ_i measures the deviation of y from the appropriate expected value. As before, β_0 and β_1 are regression coefficients, and x_i represents the age of the ith individual. The disturbance term, ϵ_i, is often called an *error* term although in many cases it may not have anything to do with measurement error. In the present example there would still have to be a disturbance term in the model even if one knew that all measurement errors had been eliminated. The term ϵ merely represents that part of the response variable y that is not accounted for by the explanatory variable x. Equation (7.22) gives a simple model for the expected frequencies in a two-dimensional contingency table. The model for the observed counts, n, would be

$$n_{ij} = m_{ij} + \epsilon_{ij} \tag{7.32}$$

where m_{ij} is the expected frequency for the ijth cell and ϵ_{ij} the random deviation from that expected frequency. Note that, although (7.22) is a multiplicative model, the random disturbance term in (7.32) is *added* to the corresponding expected value. A similar random deviation term could also be added to the expected proportion in equation (7.30). Random deviation terms could be introduced in other ways but all of the models discussed in this and the next three chapters will have the same general structure. That is,

$$\text{observed response} = \text{expected response} + \text{random deviation} \tag{7.33}$$

This is a major characteristic of what is called a *generalized linear model* (Nelder and Wedderburn 1972). A second characteristic is that the expected response, *or some transformation of it*, is a linear combination of the form

$$\eta = \beta_0 + \sum_{j=1}^{p} \beta_j x_j \tag{7.34}$$

The term η is called the *linear predictor* for the model. The required transformation of the expected response is called the *link function*. To clarify this, consider the examples introduced earlier in this chapter. For equation (7.14)

$$y = \eta \quad \text{(the identity link)} \tag{7.35}$$

For equation (7.25)

$$\ln(m) = \eta \quad \text{(the logarithmic link)} \tag{7.36}$$

Finally, for equation (7.31)

$$\ln\left(\frac{p}{1-p}\right) = \eta \quad \text{(the logistic link)} \tag{7.37}$$

These three link functions are the ones most commonly used in the behavioural sciences, but there are other possibilities (see for example, McCullagh and Nelder, 1989). The problem of fitting a generalized linear model is one of estimation of parameter values given a sample of observations. Then follows the question of assessing the goodness-of-fit of the model to the data. Before we can proceed to the estimation of parameters and the assessment of fit, however, we must consider how the random deviation terms in equation (7.34) might have arisen. This is done by specifying the *probability distribution* of the *observations*. Commonly specified probability distribution are

(1)   the normal distribution (in multiple regression and analysis of variance);
(2)   the Poisson distribution (counts in a contingency table);
(3)   the binomal distribution (proportions predicted by a linear-logistic model).

Note that the choice of a particular probability distribution is usually associated with a particular link function, but not always. Here we have given three combinations:

(1)   Identity link — normally distributed observations;
(2)   Logarithmic link — Poisson distributed observations;
(3)   Logistic link — binomial distributed observations.

Note that there is one possible source of confusion that arises from the common habits of referring to the probability distribution of the observations as if this were the same as the probability distribution of the random deviation terms or errors. This can be simply illustrated by reference to the counts in a contingency table. The counts may be Poisson variates, but the random deviations cannot be. The random deviations can be negative, but a Poisson variate cannot. In this text the phrase 'error distribution' will often be used as shorthand but it should be remembered that its exact meaning is 'probability distribution of the observations'.

The formation of linear models in terms of choice of error distributions and appropriate link functions was suggested by Nelder and Wedderburn (1972) who also give details of how such models can be fitted to data using maximum likelihood methods. Details of the general fitting procedure are beyond the scope of this book, but some comments on fitting specific models will be found in Chapters 8 to 10. A detailed account of generalised linear models is provided by McCullagh and Nelder (1989).

An important feature of the modelling process is to access how many parameters are required in the model to provide an adequate description of the data. Since a smaller number of parameters means easier interpretation, our aim will be to find the simplest model which represents the data adequately. For this purpose Nelder and Wedderburn (1972) suggest a goodness-of-fit criterion based on the maximised value of the log-likelihood and known as the *deviance*. For normally distributed observations the deviance is related to the familiar sums of squares met in regression and the analysis of variance as we will see in Chapter 8. For contingency table data the deviance is much like the Pearson chi-squared statistic, which we will consider in Chapter 9.

A number of special models need to be noted, the first of which is one containing as many parameters as observations. This is known as the *full or saturated* model and reproduces the data exactly, but without any simplification of interpretation. Furthest removed from the saturated model is the *null model* which proposes a common value for all observations; this is, of course, a very simple model, and one that in most cases will not adequately represent the structure of the data. Also of importance in some cases, particularly in dealing with contingency tables, is the *minimal model*, which arises in those situations where certain parameters *must* be included in any sensible model. For example, in a two-way contingency table, equation (7.24) would be the minimal model, since it allows for differences in cell frequencies that are attributable simple to differences in row and column marginal totals.

A point that should be noted here is that some of the models, as formulated in this chapter, are *over-parameterised*. For example, the model described by Equation (7.12) contains six parameters, μ, α_1, α_2, α_3, α_4, and α_5. Since, however, we have observations on only five social classes, we can estimate a maximum of five parameters in any model, we cannot produce independent estimates of all the factor-level effects in the presence of a parameter μ representing the grand mean. This is analogous to attempts to solve a set of independent simultaneous equations when there are more unknowns than equations. Possible methods of dealing with this problem are discussed in the next chapter.

Nelder and Wedderburn's very general formulation of linear models leads naturally to the idea of a general computer package for fitting such models, estimating their goodness of fit and so on. Such a program has been developed by Nelder and his co-workers and is known as GLIM (generalised linear interactive modelling) — see Appendix A for details. By allowing a number of possible error distributions, and a variety of link functions, a large number of models can be accommodated in the general framework. Since the program is an interactive one, it allows the possibility of exploring a large number of alternative models reasonably quickly, and this can be of considerable advantage in the model-building process outlined in Chapter 1. An introduction to GLIM can be found in Healy (1988) and a much more detailed account in Aitkin *et al.* (1989).

Exercises

7.1   By summing equation (7.23) over i, over j, and finally over i and j, show that it may be expressed in the form of equation (7.24). How could the model in (7.24) be extended to allow for a possible interaction between the row and column variables forming the table?

7.2   In a two-way analysis of variance design with r levels of one factor and c levels of the other factor, and n observations per cell, the usual model considered is

$$y_{ijk} = \mu + \alpha_i + \beta_j + (\alpha\beta)_{ij} + \epsilon_{ijk}$$

(see, for example, Winer 1971). How could this model be coded in terms of dummy variables?

7.3   In a three-way contingency table, what would be the usual form of the minimal model and the saturated model?

8

Regression and The Analysis of Variance

8.1 Introduction

Regression and the analysis of variance are both topics which are familiar to most social and behavioural scientists. They are covered in all but the simplest statistical textbooks, and the literature surrounding them is enormous. Consequently, it would be impossible to provide a comprehensive account of them in a single chapter such as this. What we shall attempt instead, however, is to show the essential equivalence of the two techniques and how each may be regarded as an aid to model building for particular types of data. (For a much more comprehensive treatment of the two topics readers are referred to Draper and Smith, 1981, Chatterjee and Price, 1977, Finn, 1974, and Rawlings, 1988. In addition a lot may be learnt from the very stimulating book by Mosteller and Tukey, 1977.)

8.2 Least Squares Estimation for Regression and Analysis of Variance (ANOVA) Models

Suppose we have several measurements for each of a sample of n individuals. For the ith individual there is a measurement of a continuous response variable, y_i, and the values of p explanatory variables $x_{i1}, x_{i2}, \ldots, x_{ip}$. (Some of the latter may be dummy variables.) To investigate the effect of the explanatory variables on the response variable we shall assume a linear model of the form

$$y_i = \mathbf{x}_i'\boldsymbol{\beta} + \epsilon_i, \quad i = 1, \ldots, n \tag{8.1}$$

where $\mathbf{x}_i' = [1, x_{i1}, x_{i2}, \ldots, x_{ip}]$ and $\boldsymbol{\beta}' = [\beta_0, \beta_1, \ldots, \beta_p]$ the vector of the parameters that require estimating. If we assume that the random disturbance terms, ϵ_i, are normally distributed with constant variance, σ^2, then the method of maximum likelihood estimation mentioned in Chapter 7 is equivalent to the more familiar least squares estimation procedure. Let the vector of parameter estimates be represented by $\hat{\boldsymbol{\beta}}' = [\hat{\beta}_0, \hat{\beta}_1, \ldots \hat{\beta}_p]$. Then it follows that

$$e_i = y_i - \mathbf{x}_i'\hat{\boldsymbol{\beta}} \tag{8.2}$$

where e_i is the *residual* for the ith observation. Note that e_i is not the same as the true error ϵ_i. In least squares estimation we search for the values of the parameter estimates that jointly minimise the sum of the squared residuals, $\sum_{i=1}^{n} e_i^2$. If we introduce the vector $y' = [y_1, y_2, \ldots, y_n]$ and an $n \times (p+1)$ matrix X, given by

$$X = \begin{pmatrix} 1 & x_{11} & x_{12} & \cdots & x_{1p} \\ 1 & x_{21} & x_{22} & \cdots & x_{2p} \\ \vdots & \vdots & \vdots & \cdots & \vdots \\ 1 & x_{n1} & x_{n2} & \cdots & x_{np} \end{pmatrix}$$

then the model corresponding to equation (8.1) may be written

$$y = X\beta + \epsilon \tag{8.3}$$

where $\epsilon' = [\epsilon_1, \epsilon_2, \ldots, \epsilon_n]$. To obtain least squares estimates one needs to minimise $e'e$ with respect to the parameter values, where $e' = [e_1, e_2, \ldots, e_n]$ is a vector of residual values. This procedure leads to the so-called *normal equations* for the parameter estimates. These are:

$$X'X\hat{\beta} = X'y \tag{8.4}$$

(Details of the derivation of (8.4) are given in Draper and Smith, 1981.)

Two situations involving the solution of the normal equations can be distinguished. The first and simplest is where the matrix $X'X$ is invertible, leading directly to $\hat{\beta}$. The second is where too many parameters have been introduced by the specification of the model and accordingly $X'X$ cannot be inverted to yield unique estimates. We shall consider the two cases in turn under Sections 8.3 and 8.4.

8.3   Multiple Regression Models

The investigator usually wishes to use the techniques of multiple regression in situations when the explanatory variables are quantitative, and the matrix $X'X$ contains the sums of squares and cross-products derived from the measurements of these variables. Here the solution of the normal equations is straightforward, and leads to

$$\hat{\beta} = (X'X)^{-1}X'y \tag{8.5}$$

A number of numerical examples are discussed in Section 8.6.

Under certain assumptions the estimates of the regression coefficients obtained from equation (8.4) are the 'best' estimates in the sense that, of all the estimates that are unbiased (their expected values being equal to the true parameter values), they have minimum variance. This also applies to the estimates for the ANOVA models of the next section. In both cases the estimates are generally known as ordinary least squares (OLS) estimates, and the most important assumptions necessary for them to be optimal are as follows:

(1)   each disturbance term has the same variance;
(2)   the disturbance terms are uncorrelated with each other;
(3)   the disturbances are statistically independent of the explanatory variables;
(4)   the values of the explanatory variables are known without error.

It is clear that these assumptions are unlikely to be met in all data sets arising in the behavioural or social sciences. Often, however, small violations of the assumptions do not lead to major problems. Still, it is important to try to detect any departures from assumptions that do occur, and, if possible, to adjust estimation procedures accordingly. (These topics will be discussed later in this chapter.)

A further assumption that ought to be satisfied is that the values of the explanatory variables be non-stochastic; that is, their values should be fixed or selected in advance. Clearly this is very rarely the case in the social sciences, so all inferences are assumed in practice to be *conditional* on the values of the explanatory variables observed.

Interpretation of a multiple regression equation is very dependent on the implicit assumption that the explanatory variables are not strongly inter-related. The regression coefficients may be interpreted as a measure of the change in the response variable when the corresponding explanatory variable is increased by one unit and all other explanatory variables are held constant. Such an interpretation would no longer be valid in the presence of strong linear relationships amongst the explanatory variables, simply because in such situations it is obviously impossible to change one variable whilst holding all others constant.

When there is a complete absence of linear relationships among the explanatory variables, they are said to be orthogonal. In most applications this condition will not hold, but small departures from it are unlikely to affect the analysis at all seriously. However, if two or more of the explanatory variables are highly correlated then the statistical significance of these variables will be very dependent on the order in which they are added to the model, and the resulting parameter estimates will be very unstable.

The condition of severe non-orthogonality of the explanatory variables is often referred to as the problem of *collinear data* and, in practice, we would first like to be able to detect collinearity and then, if possible, to cope with it. One method of detection is to apply principal components analysis to the covariance matrix of the explanatory variables (see Chapter 4). Collinearity is indicated by one or more very small eigenvalues. It is also possible to watch out for instability in parameter estimates as differing models are fitted and also, of course visually examine the correlation matrix of the explanatory variables.

Once detected, the problem of collinearity can be dealt with in a variety of ways. One involves regressing the response variable on a number of the important principal components. These are also useful in identifying those linear combinations of the regression coefficients that can be accurately estimated (see Chatterjee and Price, 1977, Chapter 7). A further possibility is *ridge regression*. This involves an adaptation of least squares estimation, and gives parameter estimates which are biased, but which in general will be 'closer' to the true parameter values than those obtained by the use of OLS in circumstances of severe departures from orthogonality. The so-called ridge estimator of the regression coefficients, $\hat{\boldsymbol{\beta}}^*$, is obtained by solving the equation

$$(\mathbf{X}'\mathbf{X} + k\mathbf{I})\hat{\boldsymbol{\beta}}^* = \mathbf{X}'\mathbf{y} \tag{8.6}$$

to give

$$\hat{\beta}^* = (X'X + kI)^{-1}X'y \qquad (8.7)$$

where k is a constant and I is the identity matrix. The essential parameter that distinguishes ridge regression from OLS is k. When $k = 0$, $\hat{\beta}^*$ is simply the usual least squares estimate. As k increases from zero, the bias of the estimates increases, and as it continues to increase without limit, the regression estimates all tend toward zero. Hoerl and Kennard (1970) have shown that there is a positive value of k for which the ridge estimate will be resistant to small changes in the data. In practice a range of values of k (between 0 and 1) are explored, and the estimates of the regression coefficients plotted against k. The resulting graph is known as a *ridge trace*, and may be used to select an appropriate value for k.

To summarise, the rationale behind ridge regression is essentially to overcome the problems involved with the inverse of the matrix $X'X$, which in situations where collinearity is present, is ill-conditioned. Adding a small constant to the diagonal elements of $X'X$ leads to biased estimates, but improves the properties of the inverse and allows more stable estimates of the regression coefficients. Examples of the use of ridge regression are given in Chatterjee and Price (1977).

Returning to OLS estimates, the predicted values for y are given by

$$\hat{y} = X\hat{\beta}$$
$$= X(X'X)^{-1}X'y$$
$$= Vy \qquad (8.8)$$

where $V = X(X'X)^{-1}X'$ is called the *'hat' matrix*. For the ith data value we have

$$\hat{y}_i = v_{ii}y_i + \sum_{j \neq i} v_{ij}y_j \qquad (8.9)$$

where v_{ij} is the ijth entry of the hat matrix, V. It is clear from equation (8.9) that $\hat{y}_i$ is a weighted average of all of the y's. The weight v_{ii} is a measure of the influence of the observation y_i on the corresponding predicted value, this influence being relative to the values of the other v_{ij} terms. Hoaglin and Welsh (1978) have used the term *leverage* for v_{ii}. The importance of the hat matrix V will be discussed further in Sections 8.8 and 8.9. Here we will briefly mention two of its important mathematical properties. The matrix V is symmetric ($V' = V$) and idempotent ($V^2 = V$). From these two properties it follows that trace $(V) = \text{rank}(V) = p + 1$ and $\sum_{j=1}^{n} v_{ij}^2 = v_{ii}$. Other properties of the hat matrix can be found in Cook and Weisberg (1982).

8.4   ANOVA Models

Let us reconsider the example introduced in Chapter 7, involving the effect of social class on the number of visits to a physician, and suppose that we have a single individual from each class and a corresponding record of the number of visits. If social class is indicated by the use of five dummy variables (see Table 7.1), the linear model equivalent to equation (8.3) for these five individuals is

$$
\mathbf{y} =
\begin{pmatrix}
1 & 1 & 0 & 0 & 0 & 0 \\
1 & 0 & 1 & 0 & 0 & 0 \\
1 & 0 & 0 & 1 & 0 & 0 \\
1 & 0 & 0 & 0 & 1 & 0 \\
1 & 0 & 0 & 0 & 0 & 1
\end{pmatrix}
\begin{pmatrix}
\mu \\ \alpha_1 \\ \alpha_2 \\ \alpha_3 \\ \alpha_4 \\ \alpha_5
\end{pmatrix}
+ \mathbf{e}
\tag{8.10}
$$

The normal equations for the estimation of the parameters are as in equation (8.3). These normal equations, however, cannot be solved to yield unique estimates. This problem arises because the model, as formulated, requires the estimation of six parameters from only five observations, one from each social class. The problem would remain even if we were to increase the number of observations in each of the social classes; here one would be attempting to estimate the six parameters from the five class means.

To illustrate the above points, consider the following set of hypothetical records. Let the number of visits recorded for a single individual from social class one be 5. Similarly let the count for single individuals from each of the other social classes be 7, 9, 18 and 20, respectively. The normal equations are as follows:

$$
\begin{pmatrix}
5 & 1 & 1 & 1 & 1 & 1 \\
1 & 1 & 0 & 0 & 0 & 0 \\
1 & 0 & 1 & 0 & 0 & 0 \\
1 & 0 & 0 & 1 & 0 & 0 \\
1 & 0 & 0 & 0 & 1 & 0 \\
1 & 0 & 0 & 0 & 0 & 1
\end{pmatrix}
\begin{pmatrix}
\mu \\ \alpha_1 \\ \alpha_2 \\ \alpha_3 \\ \alpha_4 \\ \alpha_5
\end{pmatrix}
=
\begin{pmatrix}
59 \\ 5 \\ 7 \\ 9 \\ 18 \\ 20
\end{pmatrix}
\tag{8.11}
$$

that is,

$$
\begin{aligned}
5\mu + \alpha_1 + \alpha_2 + \alpha_3 + \alpha_4 + \alpha_5 &= 59 \\
\mu + \alpha_1 &= 5 \\
\mu + \alpha_2 &= 7 \\
\mu + \alpha_3 &= 9 \\
\mu + \alpha_4 &= 18 \\
\mu + \alpha_5 &= 20
\end{aligned}
\tag{8.12}
$$

Since the sum of the last five rows or equations is equal to the first there are infinitely many solutions. In other words, the matrix $\mathbf{X}$ of (8.10) does not have full column rank, and, consequently, the matrix $\mathbf{X'X}$ in (8.11) does not have a simple inverse. Examples of possible solutions to these equations are given in Table 8.1. Clearly, as they stand, they are of very little use to the investigator. We will now introduce a number of ways in which this problem can be overcome.

One approach is to arbitrarily set one of the parameters to zero. One could, for example, let $\mu = 0$, or alternatively, let $\alpha_1 = 0$. In the latter case the αs to be estimated would then be measures of deviation from the number of visits for the individual in social class 1. This type of constraint on the value of one of the parameters might appear to be rather arbitrary, but in some situations, for example when one of the groups could be considered as a control, it would be thought quite reasonable. Whatever the type of contraint introduced, however,

Table 8.1 Four solutions to matrix equation (8.8)

Parameter	Solution 1	2	3	4
μ	10	5	1	1000
α_1	-5	0	4	-995
α_2	-3	2	6	-993
α_3	-1	4	8	-991
α_4	8	13	17	-982
α_5	10	15	19	-980

it does enable one to obtain unique estimates for the other parameters. That is, the matrix $\mathbf{X'X}$ is now invertible. The method of constraint adopted in most elementary treatments on the analysis of variance is to let the sum of the αs be zero (that is $\sum_i \alpha_i = 0$). In this case, equation (8.7) would be rewritten as

$$\mathbf{y} = \begin{pmatrix} 1 & 1 & 0 & 0 & 0 \\ 1 & 0 & 1 & 0 & 0 \\ 1 & 0 & 0 & 1 & 0 \\ 1 & 0 & 0 & 0 & 1 \\ 1 & -1 & -1 & -1 & -1 \end{pmatrix} \begin{pmatrix} \mu \\ \alpha_1 \\ \alpha_2 \\ \alpha_3 \\ \alpha_4 \end{pmatrix} + \epsilon \qquad (8.13)$$

(α_5 has been replaced by $-\alpha_1 - \alpha_2 - \alpha_3 - \alpha_4$.) Since $\mathbf{X'X}$ is now invertible the normal equations may be solved directly to give estimates of μ, α_1, α_2, α_3, and α_4.

The computer package SPSS (and several others) uses this method, but GLIM sets the equivalent of α_1 to zero. In this and the following chapters we will usually present analyses that have been carried out using GLIM, and the reader should make certain that the way of specifying the model is understood. Another approach to solving the problem of having matrices of deficient rank is to consider certain contrasts or linear functions of parameters that are invariant to whatever solution of the normal equations is considered. To illustrate this concept, consider the linear function 1 given by

$$l_1 = \alpha_1 - \alpha_2 \qquad (8.14)$$

For each of the solutions in Table 8.1, l_1 takes the value -2, and this will be so for all the infinitely many solutions of the normal equations (8.8). Again, consider the function, l_2, given by

$$l_2 = \alpha_1 - \frac{1}{4}(\alpha_2 + \alpha_3 + \alpha_4 + \alpha_5) \qquad (8.15)$$

This always takes the value -8.5.

But do such invariant linear functions provide useful information? To answer this question consider again l_1 and l_2; the first is a measure of the difference between the effects of social class 1 and social class 2, and the second is a measure of the difference between the effect of social class 1 and the average effect of the other four levels. Both of these might be of specific interest in particular situations.

Table 8.2 Coding of social classes in terms of binary dummary variables, leading to direct solution of normal equations

Social class	Dummy variables			
	x_1	x_2	x_3	x_4
1	0	0	0	0
2	1	0	0	0
3	1	1	0	0
4	1	1	1	0
5	1	1	1	1

The terms l_1 and l_2 are, of course, only two of many such linear functions of the parameters which have the property of being invariant to whatever solution of the normal equation is considered. Because of their invariance property they are the only functions which can be of interest, so far as the estimation of the parameters of a linear model is concerned. The question as to which particular linear functions have the invariance property is discussed by Searle (1971), Chapter 5; details of the reparameterisation of the models in terms of these linear functions is given by Finn (1974), Chapter 7. Essentially, however, the procedure is equivalent to coding a qualitative variable with k categories (in this example, social class has 5 categories) in terms of only $k-1$ dummy variables. This is always possible, and Table 8.2 shows such a coding for social class. The model for the effect of social class on visits to a physician can now be rewritten as

$$y = \beta_0 + \beta_1 x_1 + \beta_2 x_2 + \beta_3 x_3 + \beta_4 x_4 \tag{8.16}$$

so that equation (8.10) is replaced by

$$\mathbf{y} = \begin{pmatrix} 1 & 0 & 0 & 0 & 0 \\ 1 & 1 & 0 & 0 & 0 \\ 1 & 0 & 1 & 0 & 0 \\ 1 & 0 & 0 & 1 & 0 \\ 1 & 0 & 0 & 0 & 1 \end{pmatrix} \begin{pmatrix} \beta_0 \\ \beta_1 \\ \beta_2 \\ \beta_3 \\ \beta_4 \end{pmatrix} + \epsilon \tag{8.17}$$

The βs are now clearly estimable and they are, in fact, linear combinations of the parameters in (8.7): $\beta_0 = \mu + \alpha_1$, $\beta_1 = \alpha_2 - \alpha_1$, $\beta_2 = \alpha_3 - \alpha_2$, $\beta_3 = \alpha_4 - \alpha_3$, $\beta_4 = \alpha_5 - \alpha_4$. In general, an ANOVA model such as (8.4) can always be re-specified as a straightforward multiple regression model such as (8.1). When the investigator has recorded both categorical and quantitative explanatory variables the model can be specified either in terms of the parameters of the typical ANOVA model (as in the traditional *analysis of covariance*), or dummy variables can be created so that the situation is an extension or a special case of multiple regression. It is simply a matter of taste; computer packages such as GLIM can handle either method of model specification.

8.5   Testing the Fit of a Model

After fitting a model to a set of data it is necessary to assess the adequacy of the fit. To construct a significance test for this purpose we first have

Table 8.3   Analysis of variance to test the fit of the model $\mathbf{y} = \mathbf{X}\boldsymbol{\beta} + \varepsilon$

Source of variation	d.f.	Sum of squares	Mean square
Mean	1	$n\bar{y}^{-2}$	
Model after fitting mean (i.e. explanatory variables)	$r-1$	$ESS = \hat{\boldsymbol{\beta}}\mathbf{X}'\mathbf{y} - n\bar{y}^{-2}$	$EMS = ESS/(r-1)$
Residual (due to random disturbance terms)	$n-r$	$RSS = \mathbf{y}'\mathbf{y} - \hat{\boldsymbol{\beta}}\mathbf{X}'\mathbf{y}$	$RMS = RSS/n-r$
Total	n	$TSS = \mathbf{y}'\mathbf{y}$	

to assume a particular probability distribution for the random disturbance terms. In this chapter we will suppose that they are normally distributed, with constant variance. A test of the overall fit of the model can be obtained by the decomposition of the total variation in the response variable into that corresponding to variation accounted for by the model and the variation caused by random deviations from the model. The sums of squares involved in this exercise, along with their degrees of freedom, are defined in Table 8.3 (See Searle, 1971, Chapters 3 and 5, for further details). The terms in this table apply to models both of full rank and not of full rank. For the former, $r = p + 1$, and for the latter the value will depend upon the number of independent parameters that can be estimated. For example, in the social class example, r takes the value 5. If there is no relationship between the response variable and the explanatory variables, i.e. $\beta_1 = \beta_2 = \cdots = \beta_p = 0$, then

$$F = EMS/RMS \qquad (8.18)$$

will have an F-distribution with $r - 1$ and $n - r$ degrees of freedom. Of course, such a very general test is unlikely to be of much help in practice. More specific tests for identifying important explanatory variables appear later in this section.

From Table 8.3 we see that the proportion of variation in the response variable accounted for by the explanatory variables is

$$P = (TSS - RSS)/TSS \qquad (8.19)$$

that is,

$$P = 1 - RSS/TSS \qquad (8.20)$$

It is straightforward to show that P is equal to the square of the Pearson product moment correlation coefficient between the response variable, and its value as predicted by the explanatory variables, that is, $\hat{y}_i$ given by

$$\hat{y}_i = \mathbf{x}_i'\hat{\boldsymbol{\beta}} \qquad (8.21)$$

Pearson's correlation coefficient for y and $\hat{y}$ is known as the *multiple correlation coefficient*, and usually designated by the letter R. The equation $R^2 = P$ is a useful index of the goodness of fit of a regression model, but there are pitfalls in its uncritical use, as we shall illustrate later.

The F-test outlined above, for the hypothesis that none of the explanatory variables affects the response variable, is likely to be significant in most applications, leading to a rejection of this very general hypothesis. On its own this is not of great interest since the investigator is more often concerned with assessing whether a subset of the explanatory variables provides an adequate explanation for the variation in the response variable. If a small number of the explanatory variables produces a linear model which fits the data only marginally worse than a larger set, the investigator would usually be happier with the former since it would provide a more parsimonious model, easing the, at times, difficult problem of interpretation.

In part, the solution to this problem lies in computing the various sums of squares already met earlier in this section. For example, consider first the fitting of the following model to a set of data:

$$y = \beta_0 + \beta_1 x_1 + \cdots + \beta_p x_p \tag{8.22}$$

(We shall assume here that we are dealing with a model of full rank, although the procedure to be outlined may also be applied to other models, as we shall see in Section 8.6.) Sums of squares due to the explanatory variables and a residual sum of squares can be calculated as outlined in Table 8.2 and from these an F-statistic may be constructed to test the fit of the model. A significant F implies that we should reject the hypothesis that the regression coefficients $\beta_1, \beta_2, \ldots, \beta_p$ are all equal to zero. It does not, however, indicate whether all are non-zero, or whether all explanatory variables are needed in the model.

Now let the term $\beta_p x_p$ be deleted from the model in equation (8.22). Again the sum of squares due to the modified model can be calculated, as can the difference between this sum of squares and the regression sum of squares due to the original model. On the assumption that $\beta_p = 0$, this difference is distributed as χ^2 with a single degree of freedom. Consequently an F-statistic with 1 and $n - p - 1$ degrees of freedom can be obtained by dividing this difference sum of squares by the original residual sum of squares after fitting equation (8.19). This can then be used to test whether $\beta_p = 0$ or, in other words, whether explanatory variable x_p contributes significantly to the model, over and above variables $x_1, \ldots, x_{p-1}$. We could have removed any of the other $p - 1$ explanatory variables instead of x_p and tested their significance accordingly. Note that in each case the significance of the deleted term is conditional on the presence of the other $p - 1$ terms; that is, the importance of the deleted term is being assessed by a conditional significance test. An alternative approach is to add an extra variable and test its significance. This procedure can be extended in an obvious way to test whether a subset of regression coefficients is equal to zero (see Draper and Smith, 1981, for details).

Using this method of conditional significance testing, all possible combinations of explanatory variables could be tried, and the 'best' selected-with this taken to mean the simplest model (fewest explanatory variables) compatible with the data. The number of possible models to be examined is $2^p - 1$, and, if the set of possible explanatory variables is large, this method of model building becomes practically impossible to use, despite the development of a number of algorithms designed to reduce the computational difficulties (see, for example, Hand, 1981a). Consequently we are led to consider methods that do not need to take into account every possible model. The most popular of these are ones involving some kind of stepwise fitting procedure. There are basically two approaches to stepwise selection of variables, forward selection and backward elimination. We shall now consider each of these in turn.

The procedure of forward selection of variables starts by fitting just a constant term (the mean) to the observations, i.e. the parameter β_0. Then each of the possible variables is added to the model in turn, and the most significant, provided that the significance level is below some predetermined level, is selected for inclusion. Each of the remaining variables is then added in turn, and again the most significant selected for inclusion. This procedure is repeated until there is no further significant improvement in the fit of the model. The procedure can be modified in several ways, for example, by fitting pairs of variables sequentially rather than trying each separately. In addition, the variables already included in the model can be considered again in turn at each of the stages in the process to see if they still make a significant contribution to the fit of the model. If not, they can be deleted.

Backward elimination of variables starts with the most complicated of the possible models; that is, it starts with a model containing all of the possible explanatory variables. Then each variable is deleted in turn, and the least significant left out, provided that the significance level is above some predetermined level. The process is repeated until the simplest model that is compatible with the data is obtained. Often the resulting model will be identical to that found by forward selection of variables, but this is not necessarily so. It should be stressed that both of these procedures or algorithms should be used with care. They are not foolproof. They should be used as aids in the model-building process, and, wherever possible, they should be used in conjunction with more intuitive model selection procedures based on the theoretical knowledge of the investigator. We will not discuss these methods in any detail here, but an example of the use of forward selection and backward elimination methods is given in the next section and in Chapter 9.

Other interesting methods which have been developed for selecting subsets of explanatory variables are discussed by Mallows (1973) and Allen (1971).

8.6   Numerical Examples

To illustrate some of the topics discussed in the previous sections we shall begin by considering the hypothetical data on visits to a physician given in Table 8.4. The results of fitting a number of models to these data appear in Table 8.5. In part (a) the sum of squares due to the addition of the variable weight to the model is $23.24-16.35=6.89$ with one degree of freedom. Note that this is considerably less than that due to weight in part (b), the reason being the correlation between age and weight. In part (b) the sum of squares due to the addition of age to the model is $23.24-16.51=6.73$, again with 1 degree of freedom. It will be left as an exercise for the reader (see Exercise 8.2) to construct appropriate F-tests to assess the fit of these models.

Now let us examine the model fitting procedure for examples of the analysis of variance (ANOVA) type. Consider first a design with two qualitative factors A and B, and with an equal number of observations in each cell. Suppose we fit the following series of models:

$$y = \text{mean},$$

$$y = \text{mean} + \text{effect of factor A},$$

$$y = \text{mean} + \text{effect of factor A} + \text{effect of factor B},$$

Table 8.4 Hypothetical data on number of visits to a physician

Individual	Age (yr)	Weight (kg)	Number of visits in one year
1	15	45	2
2	30	80	4
3	35	70	2
4	40	90	6
5	43	75	3
6	48	130	5
7	65	75	3
8	67	120	8
9	72	90	5
10	80	75	7

Table 8.5 Fitting linear regression models for the number of visits to a physician shown in Table 8.4

(a) *Fitting age first*

Model	s.s. (d.f.)	Residual s.s. (d.f.)
Visits $= \beta_0$	38·50 (9)	
Visits $= \beta_0 + \beta_1 \times$ Age	16·35 (1)	22·15 (8)
Visits $= \beta_0 + \beta_1 \times$ Age $+ \beta_2 \times$ Weight	23·24 (2)	15·26 (7)

(b) *Fitting weight first*

Visits $= \beta_0$	38·50 (9)	
Visits $= \beta_0 + \beta_1 \times$ Weight	16·51 (1)	21·99 (8)
Visits $= \beta_0 + \beta_1 \times$ Weight $+ \beta_2 \times$ Age	23·24 (2)	15·26 (7)

$$y = \text{mean} + \text{effect of factor A} + \text{effect of factor B}$$
$$+ \text{AB interaction effect.}$$

By employing the procedure outlined in the previous examples we could find a sum of squares A given the mean, for B given A and the effect of the mean, and for AB given the effects of the mean and A and B. Here, however, the order of adding terms to the model is irrelevant since the design is orthogonal. Consequently terms such as sum of squares due to B given A and the mean, and the sum of squares due to B given the mean, will be the same. The total sum of squares of the response variable may be partitioned into non-overlapping components corresponding to the effect of factor A, the effect of factor B, the AB interaction effect and a residual term. These will be the usual sums of squares in the analysis of variance table for such a design.

If we now consider the same simple design but assume that there are different numbers of observations in each cell the analysis becomes more complex, since the order of entering effects into the model now becomes of importance. We

Table 8.6 Example of a non-orthogonal analysis of variance design

	B_1	B_2
A_1	10	15
	5	20
		25
		30
A_2	5	10
	8	12
	8	

Table 8.7 Results of fitting various models to data of Table 8.6

(I) *Fitting A, followed by B, followed by AB interaction*		d.f.
Sum of squares due to A given the mean	= 216·0	1
Sum of squares due to B given A and the mean	= 242·8	1
Sum of squares due to AB given A, B and the mean	= 76·4	1
Residual sum of squares	= 145·5	7

(II) *Fitting B, followed by A, followed by AB interaction*		
Sum of squares due to B given the mean	= 358·6	1
Sum of squares due to A given B and the mean	= 100·2	1
Sum of squares due to AB given A, B and the mean	= 76·4	1
Residual sum of squares	= 145·5	7

can illustrate this with the small set of data shown in Table 8.6. The results of fitting various models appear in Table 8.7.

Because of the different numbers of observations in each cell the order of entering parameters into the models is of importance. We see, for example, that the sum of squares for B given the mean, and that for B given A and the mean are not the same. The sums of squares in these tables were obtained from the 'deviance' goodness-of-fit criterion of the GLIM program, specifying an 'identity' link function and normally distributed errors. For example, the deviance for a model containing only a term for the mean is 680.7, with 10 degrees of freedom. This is equivalent to the sum of squares in the response variable remaining after the mean has been included in the model. Adding a parameter for factor A results in a deviance of 464.7 and 9 degrees of freedom. This is equivalent to the residual sum of squares after fitting the mean and A. Subtracting this from the previous deviance gives the additional sum of squares for A over that due to the mean, that is, 216.0. Other sums of squares are obtained in a similar fashion.

As a further example of a non-orthogonal design we will consider the data originally discussed by Francis (1973), in which individuals categorised by religion and by sex were scored on a six-point psychological scale. The means and number of observations in each cell appear in Table 8.8. The various sums

Table 8.8 Means and frequencies, for example from Francis (1973)

Religion	1	2	3	4	5
Sex 1	3·468	3·511	3·368	2·700	2·973
	(299)	(131)	(38)	(60)	(37)
2	3·143	2·925	3·097	2·667	2·983
	(342)	(207)	(72)	(165)	(59)

Table 8.9 Models fitted to Francis data

Source	s.s.	d.f.	Source	s.s.	d.f.
Sex	43·58	1	Religion	69·36	4
Religion given the effect of Sex	54·49	4	Sex given the effect of Religion	28·71	1
Sex × Religion given the main effects of Sex and Religion	10·25	4	Sex × Religion given the main effects of Sex and Religion	10·25	4
Residual sum of squares	2988·95	1300	Residual sum of squares	2988·95	1300

of squares associated with the fitting of various models are given in Table 8.9. These indicate that a satisfactory model for these data would contain a term for sex and a term for religion, but that an interaction term would be unnecessary. These data are particularly interesting since Francis used them to show that several computer packages for analysis of variance were failing to deal appropriately with non-orthogonal designs. The analysis of unbalanced cross-classifications can be a troublesome task, and the reader concerned with such designs is urged to consult the two excellent papers of Nelder (1977) and Aitkin (1978) and the book by Searle (1987).

Next, we shall consider a relatively complex analysis of data derived from a learning experiment on eye-blink conditioning. The aim of the experiment was to condition the subjects to react to an audio tone by blinking, and this was achieved by presenting the audio tone together with an electric shock. The extent of learning was assessed by the number of conditioned responses (eye-blinks) given in 50 acquisition trials. The results are shown in Table 8.10. There were a number of factors that were thought to influence the extent of conditioning and measurements on eight of these covariates are also given in Table 8.10. They are defined in Table 8.11. Some of them are interval- scaled, others merely have ordinal or qualitative significance. In the analysis described here, the variables APR, PSY, EXTR and CAWR were treated as quantitative covariates, and Q14, Q15, Q20 and THRE were models when the explanatory variables are a mixture of quantitative measurements and qualitative factors.

Consider now Table 8.12(a). This shows an analysis of variance for a multiple regression model incorporating all eight of the explanatory variables. The words 'by' and 'with' in the heading are used to distinguish the qualitative and quantitative variables. It has been assumed that there are no interaction

Table 8.10 *Continued*

TOTCR	APR	PSY	EXTR	CAWR	THRE	Q14	Q15	Q20
0	31	6	16	9	2	4	3	1
0	36	3	16	9	2	4	5	1
0	38	1	16	7	1	4	4	2
0	41	2	19	14	1	5	6	2
0	24	1	21	−16	1	5	4	1
0	32	0	15	14	1	4	4	1
0	31	7	14	4	2	4	3	2
1	30	8	15	14	1	5	3	1
1	35	5	4	14	2	3	4	1
1	37	8	17	9	1	6	2	1
1	31	8	7	9	2	5	2	1
1	42	5	19	14	1	4	3	1
1	23	1	19	14	2	4	3	1
1	41	5	8	−2	2	6	4	3
1	33	2	19	11	2	4	2	1
1	47	3	9	9	2	6	3	2
2	32	4	18	−18	2	4	4	3
2	35	1	11	−1	1	5	5	1
2	41	5	19	−13	1	4	5	3
2	49	4	6	−18	2	4	2	1
2	31	5	5	9	1	4	5	1
3	43	9	17	9	2	4	4	3
3	21	1	16	−2	2	6	6	1
3	37	4	2	14	2	5	6	3
3	33	2	14	14	2	4	2	1
4	49	8	20	14	1	1	4	1
4	35	9	14	−19	1	5	4	1
5	23	6	16	−11	2	4	5	1
5	45	4	15	−1	1	7	6	1
5	37	1	11	9	2	4	4	7
5	41	2	7	14	2	5	6	7
6	49	4	15	14	1	4	6	3
6	43	6	18	14	2	4	4	3
6	39	8	12	9	2	5	4	3
7	30	7	19	−19	1	4	5	2
8	26	1	16	4	1	5	5	3
9	27	3	16	−1	2	5	4	1
9	32	3	11	9	2	6	5	3
9	32	9	13	14	2	4	4	1
9	30	2	15	14	2	4	3	1
9	33	3	10	14	1	4	4	1
9	30	2	3	9	1	5	5	1
9	33	6	18	9	1	4	4	3
10	34	4	12	4	1	5	6	6
11	40	10	7	−18	2	5	5	1
11	34	6	16	9	1	6	5	6
11	56	5	5	9	2	4	4	4
12	25	4	14	14	1	4	3	2
12	36	1	19	−24	1	6	5	2
12	31	1	8	−1	1	4	4	1
12	30	12	9	9	2	5	3	3
13	28	2	13	−6	2	4	2	1
13	21	8	18	4	2	4	6	1
13	50	5	12	14	1	7	7	1
14	37	5	15	9	2	7	6	7
14	32	1	10	14	2	5	5	1
14	40	7	19	−13	1	7	5	6

Table 8.10   Eye blink conditioning data (from Freka *et al.*, 1982)

TOTCR	APR	PSY	EXTR	CAWR	THRE	Q14	Q15	Q20
14	46	4	7	2	2	4	4	4
15	30	4	8	4	2	6	6	1
15	24	13	10	9	2	5	2	4
15	44	6	11	−2	2	6	5	7
15	39	5	20	−10	1	5	4	2
17	32	4	19	2	1	4	4	1
18	30	2	11	11	2	4	2	2
19	42	4	11	9	2	5	5	4
19	41	3	20	14	1	5	4	1
21	26	2	18	14	1	4	4	6
21	34	4	2	14	2	6	2	1
22	35	0	19	14	2	5	5	3
22	29	1	11	9	1	4	4	6
23	26	7	11	14	1	5	4	6
24	39	1	16	7	1	6	6	1
25	45	0	9	8	1	6	5	3
25	31	3	19	−8	2	4	4	5
26	35	4	10	9	1	6	5	3
27	45	3	15	4	1	4	5	1
29	26	1	12	−19	1	4	4	6
30	38	5	14	14	1	6	5	5
30	35	3	20	−14	1	4	4	3
32	40	1	8	−8	1	5	6	1
33	35	3	21	−1	1	4	6	6
34	48	1	3	25	1	6	1	1
35	31	5	14	9	1	6	5	3
35	24	0	20	14	1	6	5	1
43	37	1	15	−18	1	6	5	4

Table 8.11   Definitions of variables listed in Table 8.10*

TOTCR	The total number of conditioned responses given in 50 acquisition trials
APR	Score for anxiety
PSY	Score for psychoticism
EXTR	Score for extraversion
CAWR	Contingency awareness assessment
THRE	Binary variable indicating whether the subject's threshold for a blink reflex is over a given shock level
Q14	Ordinal scale for subject's rating of the pleasantness of the shock
Q15	Ordinal scale for the subject's rating of the strength of the shock
Q20	Ordinal scale indicating frequency in which subject blinked purposely to avoid the shock

* Details of the experiment are given in the text.

terms required in the model. The significance of end of the model terms is assessed from an *F*-statistic derived from the decrease in the sum of squares explained by the model, on deletion of the appropriate term from the full model (containing all eight explanatory variables). That is, the statistical significance of these terms is assessed from a conditional *F*-test.

Table 8.12　Analysis of variance for data in Table 8.10

(a) *TOTCR BY Q14, Q15, Q20, THRE WITH APR, PSY, EXTR, CAWR*

Source of variation	Sum of squares	d.f.	Mean square	F	Significance of F
Q14	313·098	5	62·620	0·745	0·593
Q15	89·104	5	17·821	0·212	0·956
Q20	1522·960	6	253·827	3·018	0·012
THRE	472·565	1	472·565	5·618	0·021
APR	68·471	1	68·471	0·814	0·370
PSY	204·652	1	204·652	2·433	0·124
EXTR	48·366	1	48·366	0·575	0·451
CAWR	171·125	1	171·125	2·035	0·159
EXPLAINED	3964·595	21			
RESIDUAL	5382·998	64			
TOTAL	9347·593	85			

(b) *TOTCR BY Q14, Q20, THRE WITH APR, PSY, EXTR, CAWR*

Source of variation	Sum of squares	d.f.	Mean square	F	Significance of F
Q14	442·875	5	88·575	1·117	0·360
Q20	1691·270	6	281·878	3·554	0·004
THRE	569·655	1	569·655	7·183	0·009
APR	58·888	1	58·888	0·743	0·392
PSY	273·907	1	273·907	3·454	0·067
EXTR	58·765	1	58·765	0·741	0·392
CAWR	212·572	1	212·572	2·680	0·106
EXPLAINED	3875·491	16			
RESIDUAL	5472·102	69			

(c) *TOTCR BY Q14, Q20, THRE WITH PSY, EXTR, CAWR*

Source of variation	Sum of squares	d.f.	Mean square	F	Significance of F
Q14	469·857	5	93·971	1·189	0·323
Q20	1651·276	6	275·213	3·483	0·005
THRE	514·740	1	514·740	6·515	0·013
PSY	269·343	1	269·343	3·409	0·069
EXTR	30·689	1	30·689	0·388	0·535
CAWR	205·729	1	205·729	2·604	0·111
EXPLAINED	3816·602	15			
RESIDUAL	5530·991				

(d) *TOTCR BY Q14, Q20, THRE WITH PSY, CAWR*

Source of variation	Sum of squares	d.f.	Mean square	F	Significance of F
Q14	525·870	5	105·174	1·343	0·256
Q20	1681·338	6	280·223	3·577	0·004
THRE	484·058	1	484·058	6·179	0·015
PSY	268·416	1	268·416	3·427	0·068
CAWR	200·557	1	200·557	2·560	0·114
EXPLAINED	3785·913	14			
RESIDUAL	5561·680	71			

Table 8.12 *Continued*

(e) *TOTCR BY Q20, THRE WITH PSY, CAWR*

Source of variation	Sum of squares	d.f.	Mean square	F	Significance of F
Q20	1861·589	6	310·265	3·874	0·002
THRE	608·996	1	608·996	7·603	0·007
PSY	415·238	1	415·238	5·184	0·026
CAWR	237·815	1	237·815	2·969	0·089
EXPLAINED	3260·043	9			
RESIDUAL	6087·550	76			

(f) *TOTR BY Q20, THRE WITH PSY*

Source of variation	Sum of squares	d.f.	Mean square	F	Significance of F
Q20	1835·668	6	305·945	3·724	0·003
THRE	701·386	1	701·386	8·538	0·005
PSY	421·571	1	421·571	5·132	0·026
EXPLAINED	3022·228	8			
RESIDUAL	6325·365	77			

By inspecting the last column of Table 8.12(a) it can be seen that very few of the F-statistics derived in this way are statistically significant (using 0.05 as the level of significance), and that for Q15 is the least significant. Table 8.12(b) shows the repeat of the analysis of variance after removing the effect of Q15 from the model. In this analysis, the conditional F-statistics are obtained in a similar way to those in Table 8.12(a), but here the full model contains the effects of only seven explanatory variables. Note that the sum of squares explained by the model in Table 8.12(b) is less than that in Table 8.12(a), the difference being that due to Q15 in Table 8.12(a). Similarly the residual sum of squares in Table 8.12(b) has increased by the same amount. In this table the least significant F-statistics are those due to the effects of APR and EXTR. The effect of APR was deleted from the model and the analysis of variance repeated (Table 8.12(c)). By continuing this procedure eventually one arrives at a fairly simple linear model containing the effects of only three explanatory variables, Q20, THRE and PSY — see Table 8.12(f).

The above analyses illustrate the method of backward elimination. The data could also have been analysed using a forward selection procedure (see Exercise 8.6). The use of either, or both, of these methods on their own, however, might lead to unjustified conclusions about the data. Before plunging into a complicated statistical analysis of this sort we should always ask whether it is reasonable to assume that the response variable can be modelled by a linear combination of the explanatory variables, and, if so, whether the assumptions required for an analysis of variance are also justified. To answer these questions a graphical analysis of the data (including the plotting of residuals) is necessary. This is the subject of the next section (but see also Exercise 8.7).

Table 8.13   Four hypothetical data sets, each comprising 11 (x, y) pairs (from Anscombe, 1973)

Data set Variable		1–3 x	1 y	2 y	3 y	4 x	4 y
Obs. no.	1	10·0	8·04	9·14	7·46	8·0	6·58
	2	8·0	6·95	8·14	6·77	8·0	5·76
	3	13·0	7·58	8·74	12·74	8·0	7·71
	4	9·0	8·81	8·77	7·11	8·0	8·84
	5	11·0	8·33	9·26	7·81	8·0	8·47
	6	14·0	9·96	8·10	8·84	8·0	7·04
	7	6·0	7·24	6·13	6·08	8·0	5·25
	8	4·0	4·26	3·10	5·39	19·0	12·50
	9	12·0	10·84	9·13	8·15	8·0	5·56
	10	7·0	4·82	7·26	6·42	8·0	7·91
	11	5·0	5·68	4·74	5·73	8·0	6·89

8.7   Checking Assumptions by Graphical Analysis and Examination of Residuals

Consider the hypothetical data given in Table 8.13 (from Anscombe, 1973). These are four sets of results, each comprising 11 (x, y) pairs. The variable x is taken to be the explanatory variable, and we wish to examine the dependence of the response variable y on x. What can be learnt from Table 8.13? Many people, when presented with observations of this type, are inclined immediately to reach to their electronic calculator or computer terminal to determine values of correlation or regression coefficients. But often, much more can be learnt from a preliminary graphical exploration of the data. It is possible to judge from graphs which statistical model is likely to be applicable to the observations, and to assess the extent of departures of the data from any initial assumptions that are likely to be made in a formal analysis of the results.

Blindly fitting a simple regression model ($y = \beta_0 + \beta_1 x$) to the four sets of data in Table 8.13 yields 3.0 and 0.5 as the estimates of β_0 and β_1, respectively, in all cases. In each case, R^2 is found to be 0.667. If presented with these statistics, most people would visualise a graph like Figure 8.1. This, in fact, is a graphical representation of data set 1. Here the linear regression models seems to be justified. But now look at Figure 8.2. This is a similar graphical representation of data set 2. Here the simple regression model is clearly inapplicable. Instead, y has a smooth curved relation with x, possibly quadratic, and there is little residual variability. Here an R^2 of 0.667 is an underestimate of the amount of variation in the values of y that can be explained by x. Figure 8.3, which illustrates data set 3, shows another potential pitfall of fitting straight lines by least squares methods before plotting simple graphs. Here, all but one of the data points lie on a straight line (but not the one yielded by ordinary least squares estimation), and the other observation lies far from this line. Deleting the atypical or outlying observation, β_0 and β_1 are estimated to be 4.0 and 0.346 respectively.

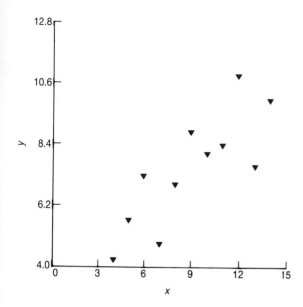

Figure 8.1

Finally, look at the scatterplot for data set 4 (Figure 8.4). Our linear model may be justified, but this depends very much on the reliability of one of the measurements. All the information about the slope of the fitted line comes from one observation, and if that observation were to be deleted from the data set the slope could not be estimated. In both data sets 3 and 4 it is one, possibly unreliable, observation that has a crucial influence on the estimation of the parameters of the regression model.

The most common way of checking the assumptions of the linear models considered in this chapter is by an examination of residual values which are defined as 'observed response less response predicted by the model'; that is, $e_i = y_i - \hat{y}_i$. The plotting of residuals is of considerable importance when assessing the applicability of a given model to a set of data, and the most useful types of plot are as follows:

(1) Plot a frequency distribution of the residuals to check for symmetry and, in particular, approximate normality.

(2) Plot the residuals as a time sequence if the order of obtaining the values of the response variable is known. This is useful for indicating whether or not the variance of the measurements is constant, and in assessing whether a term allowing for the effects of time should be included in the model.

(3) Plot the residuals against the fitted values of the response variable.

(4) Plot the residuals against any of the possible explanatory variables whether or not they have already been included in the model.

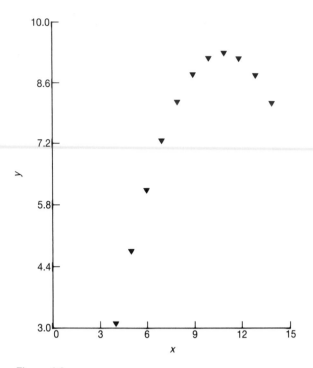

Figure 8.2

Consider Figure 8.5. These plots are typical of what one might observe when investigating the association between residuals values and time, the response variable, or one of the explanatory variables. Plot (a) is what one might expect if the fitted model were adequate. Plot (b) indicates that the assumption of constant variance is not justified and that a weighted least squares analysis is more appropriate (see Section 8.8) or that the response variable should be transformed to stabilise the variance of the measurements. Plots (c) and (d) indicate that appropriate linear or quadratic terms need to be added, respectively, to the model.

In the above discussion it has been assumed that the residuals ($e_i = y_i - \hat{y}_i$, $i = 1, 2, \ldots, n$) are an accurate reflection of the corresponding true errors (ϵ_i, $i = 1, 2, \ldots, n$). The actual relationship between e_i and ϵ_i can be found from equation (8.8). From

$$\mathbf{e} = \mathbf{y} - \hat{\mathbf{y}}$$

and equation (8.8) it follows that

$$\mathbf{e} = (\mathbf{I} - \mathbf{V})\mathbf{y} \tag{8.23}$$

Replacing $\mathbf{y}$ by $\mathbf{X}\boldsymbol{\beta} + \boldsymbol{\epsilon}$, we obtain

$$\mathbf{e} = (\mathbf{I} - \mathbf{V})\boldsymbol{\epsilon} \tag{8.24}$$

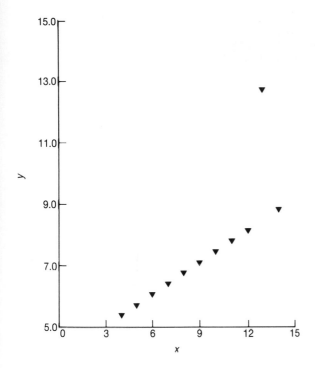

Figure 8.3

This can be expressed in scalar form as

$$e_i = \epsilon_i - \sum_{j=1}^{n} v_{ij}\epsilon_j \qquad (8.25)$$

where v_{ij} is the ijth element of **V**. If the ϵ_is are normally distributed mean 0 and variance σ^2 then the e_is will also be normally distributed with mean 0 and variance given by

$$\text{Var}(e_i) = \sigma^2 - \sum_{j=1}^{n} v_{ij}^2 \sigma^2 \qquad (8.26)$$

From the properties of the hat matrix given in Section 8.3, it follows that

$$\text{Var}(e_i) = \sigma^2(1 - v_{ii}) \qquad (8.27)$$

Equation (8.27) suggests the use of a *standardised residual* in diagnostic plots. This residual, r_i, is given by

$$r_i = \frac{e_i}{s\sqrt{1 - v_{ii}}} \qquad (8.28)$$

where s is the estimate of σ obtained from the residual mean square. This standardised residual is often referred to as a *Studentised residual*.

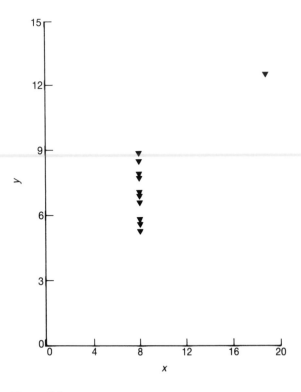

Figure 8.4

The ordinary residuals (e_i) and standardised residuals (r_i) are derived from a fit of a particular model to all of the data. Alternatively, the ith *predicted residual*, $e_{(i)}$ can be calculated from a fit of the model to the data with the ith case excluded from the analysis. If $\hat{\boldsymbol{\beta}}_{(i)}$ denotes the least squares estimates of $\boldsymbol{\beta}$ with the ith case excluded then

$$e_{(i)} = y_i - \mathbf{x}'_i\hat{\boldsymbol{\beta}}_{(i)}, \qquad i = 1, 2, \ldots, n \qquad (8.29)$$

The relationship between $e_{(i)}$ and e_i can be shown (Cook and Weisberg, 1982) to be

$$e_{(i)} = \frac{e_i}{(1 - v_{ii})} \qquad (8.30)$$

The variance of $e_{(i)}$ is given by

$$\mathrm{Var}(e_{(i)}) = \frac{1}{(1 - v_{ii})^2}\mathrm{Var}(e_i)$$

$$= \frac{\sigma^2}{(1 - v_{ii})} \qquad (8.31)$$

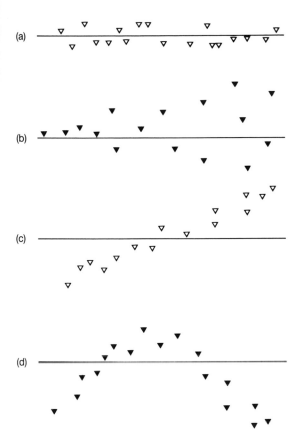

Figure 8.5

A corresponding *standardised predicted residual* can be obtained by dividing $e_{(i)}$ by an estimate of its standard error:

$$\frac{e_{(i)}}{\left(\frac{s}{\sqrt{1-v_{ii}}}\right)} = \frac{\left(\frac{e_i}{1-v_{ii}}\right)}{\left(\frac{s}{\sqrt{1-v_{ii}}}\right)}$$

$$= \frac{e_i}{s\sqrt{1-v_{ii}}}$$

$$= r_i \qquad (8.32)$$

where s, as before, is an estimate of σ based on the residual mean square from an analysis using the full data set. Alternatively, one could estimate σ from the residual mean square derived from an analysis with the ith observation missing — $s_{(i)}$. Here the standardised predicted residual, $r_{(i)}$, would be given

Table 8.14   Standardized residuals(r_i), leverage
values (v_{ii}), and Cook's D_i

(a) For data set 3 of Table 8.13

Unit(i)	$\hat{y}_i$	r_i	v_{ii}	D_i
1	8·00	−0·46	0·10	0·012
2	7·00	−0·20	0·10	0·002
3	9·50	3·00	0·24	1·393
4	7·50	−0·33	0·09	0·006
5	8·50	−0·60	0·13	0·026
6	10·00	−1·13	0·32	0·301
7	6·00	0·07	0·17	0·001
8	5·00	0·38	0·32	0·034
9	9·00	−0·76	0·17	0·060
10	6·50	−0·07	0·13	0·000
11	5·50	0·21	0·24	0·007

(b) For data set 4 of Table 8.13

Unit(i)	$\hat{y}_i$	r_i	v_{ii}	D_i
1	7·00	0·36	0·10	0·007
2	7·00	−1·06	0·10	0·062
3	7·00	0·60	0·10	0·020
4	7·00	1·57	0·10	0·137
5	7·00	1·25	0·10	0·087
6	7·00	0·03	0·10	0·000
7	7·00	−1·49	0·10	0·124
8	12·50	0·00	1·00	0·000
9	7·00	−1·23	0·10	0·084
10	7·00	0·78	0·10	0·033
11	7·00	−0·09	0·10	0·001

by

$$r_{(i)} = \frac{e_i}{s_{(i)}\sqrt{1 - v_{ii}}} \qquad (8.33)$$

The use of predicted residuals in place of ordinary residuals in diagnostic plots will tend to emphasise data points with relatively large leverage. The use of ordinary residuals will emphasise the data points with small leverage. Table 8.14 gives standardised residuals (r_i) and leverage values (v_{ii}) obtained from an analysis of data sets 3 and 4 of Table 8.13.

8.8   Detection of Influential Observations

The diagnostic methods presented in the last section are useful for checking the assumptions on which a particular analysis is based and to check general inadequacies in a model. These methods will also help in the discovery of unusual or potentially critical observations. The purpose of this section is to discuss methods of assessing the importance of these observations for the stability of the results.

One obvious approach is simply to delete each observation in turn and then repeat the analysis. In this way one can check the influence of each observation on any statistic of interest (R^2, parameter estimates, F-tests, and so on). This approach is quite straightforward but can be computationally rather tedious if the relevant sample contains many observations. An alternative strategy might involve deletion of groups of observations for the analysis rather than deletion of individuals.

Consider the influence of observation i on the parameter estimates, $\hat{\beta}$. If the estimates obtained for the data with case i deleted are $\hat{\beta}_{(i)}$, then the influence of case i can be assessed from $\hat{\beta} - \hat{\beta}_{(i)}$. Cook (1977) has suggested that a measure of this influence can be calculated using the distance measure

$$D_i = \frac{(\hat{\beta}_{(i)} - \hat{\beta})'(X'X)(\hat{\beta}_{(i)} - \hat{\beta})}{(p+1)s^2} \qquad (8.34)$$

where s^2 is the residual mean square obtained for the use of the whole sample in the analysis. Cook (1977) demonstrates that D_i can also be expressed as

$$D_i = \frac{1}{(p+1)} r_i^2 \frac{v_{ii}}{(1 - v_{ii})} \qquad (8.35)$$

Apart from the constant term $(p+1)$ this is a simple function of the standardised residual (r_i) and the leverage (v_{ii}). The advantage of (8.35) is that it is not necessary to carry out $n+1$ regressions in order to assess the influence of each of the n observations.

Various interpretations of the ratio $v_{ii}/(1 - v_{ii})$ are provided by Cook and Weisberg (1982). Cook (1977) noted that

$$\frac{v_{ii}}{(1 - v_{ii})} = \frac{\text{Var}(\hat{y}_i)}{\text{Var}(e_i)} \qquad (8.36)$$

and interpreted this ratio as a measure of sensitivity of the estimate, $\hat{\beta}$, to potentially outlying values at each data point. The term r_i^2 was interpreted as a measure of the degree to which the ith observation can be considered as an outlier from the assumed model. D_i values for data sets 3 and 4 of Anscombe (1973) are provided in Table 8.14. It will be left to the reader to interpret these results in terms of the plots given in Figures 8.3 and 8.4, respectively.

8.9   Weighted Least Squares

When an ordinary least squares (OLS) estimation is carried out all the measurements of the response variable are given equal weight; that is, they are assumed to be of equal importance in the estimation of the parameters of the required regression model. If the assumptions for linear regression hold (see Section 8.3) the resulting estimates are also the best of the possible unbiased estimates of the true parameter values. What should be done, however, if it is assumed that some of the measurements are probably less reliable or less trustworthy than the others? If this were true the OLS estimates would then not be the best. Here, we will briefly discuss methods that have been investigated to get around this problem. The emphasis of the discussion will be on methods of data exploration, mainly to give the student a better feel for the analysis of data using multiple regression; little or no mention of statistical significance testing will be made.

In the discussion of graphical exploration of data and analysis of residuals (see Sections 8.7 and 8.8) it was pointed out that possible unreliable measurements, in the form of extreme values, or outliers, are often detected. They are characterised, after OLS fitting of the required model, by corresponding extreme residuals. They might be abnormal merely by chance, or they might be the results of mistakes in the measurement or coding of the data; it is usually impossible to tell which. From the discussion of influence in Section 8.8 it should be clear that outliers can have an inordinate effect on the estimated parameter values. OLS is extremely sensitive to the presence of outliers, and one method of overcoming this sensitivity to produce a more *robust* method, is to give the more abnormal measurements less weight or importance. An extreme example of this weighting is deletion of the observation(s) for the data set. One could argue, however, that it is the outliers that are potentially more illuminating in the analysis of a set of results. For this reason it is still wise to try OLS estimation after a preliminary graphical analysis, and then, after careful examination of the residuals, proceed to delete outliers or to a weighted analysis of the results.

Consider the deviation of an observed measurement of the response variable, y_i, for its predicted or estimated value, $\hat{y}_i$. OLS minimises the sum of squares of these deviations; that is, it minimises

$$\sum_{i=1}^{n} e_i^2 = \sum_{i=1}^{n} (y_i - \hat{y}_i)^2 \tag{8.37}$$

Weighted least squares gives to each deviation or residual a weight, w_i, so that it is the following expression that is minimised:

$$\sum_{i=1}^{n} w_i (y_i - \hat{y}_i)^2 \tag{8.38}$$

That is, each measurement, y_i, is given a corresponding, weight, w_i. All we need now is a sensible choice for the w_i. If we knew the variance of each of the y_i (assuming that it might vary from one measurement to another), one obvious method of weighting would be to make the w_i inversely proportional to the corresponding variances. Here each weighted deviation can be thought of as having equal reliability or precision. Usually, however, the variances will not be known.

A modification of OLS is to minimise the sum of the predicted residuals (predicted residual sum of squares) given by

$$PRESS = \sum_{i=1}^{n} e_{(i)}^2 \tag{8.39}$$

This is the model-fitting criterion used by Anderson et al. (1972) and Allen (1974). From (8.30) it should be clear that this is a form of weighted least squares with $w_i = (1 - v_{ii})^2$. Other possibilities include the minimisation of $\sum_{i=1}^{n} r_i^2$ or $\sum_{i=1}^{n} r_{(i)}^2$ (see Cook and Weisberg, 1982).

Weighted least squares can be used in situations other than where there are suspect measurements. Sometimes it is clear from a preliminary graphical or OLS analysis that the variance of the response variable is not constant. The variance, for example, could be proportional to the fitted values of the response variable, or proportional to the square of the fitted value, and so on. An obvious way to proceed with the analysis would be to re-fit the model by minimising a function proportional respectively, either to

$$\sum_{i=1}^{n} \frac{(y_i - \hat{y}_i)^2}{y_i} \tag{8.40}$$

or to

$$\sum_{i=1}^{n} \frac{(y_i - \hat{y}_i)^2}{y_i^2} \tag{8.41}$$

Allocation of the prior weights to the observations can easily be undertaken when using a model-fitting package such as GLIM.

Consider a further example, where the linear model is to be fitted to the means of grouped observations. Suppose that the individual observations making up the groups have a common variance, say, σ^2, but that the groups comprise different numbers of observations. For group i, containing n_i observations, the variance of the mean is $\frac{\sigma^2}{n_i}$. An obvious candidate for the weight w_i for this mean is proportional to n_i. That is, we attempt to minimise a function proportional to

$$\sum_{i=1}^{n} n_i (y_i - \hat{y}_i)^2 \tag{8.42}$$

Finally, we return to the problem of coping with outliers. Consider data set 3 of Table 8.13. We have shown that the OLS estimates of β_0 and β_1 in the equation $y = \beta_0 + \beta_1 x$ are 3.0 and 0.5, respectively. If the outlying observation is discarded (that is, it is given a prior weight of zero) these estimates change to 4.0 and 0.346 respectively. These are the two extreme pairs of estimates. In the first, all of the weights are equal to 1; in the second all except the outlier are set to 1. Another possible alternative is first to find the OLS estimates, and then re-estimate the parameter values using weights that are inversely proportional to the absolute values of the corresponding residuals. But is this the best that can be done? Not necessarily, since new residuals can be calculated using these weighted least squares estimates, and a second weighted analysis carried out using weights inversely proportional to the absolute values of the second set of residuals. This cycle of estimation can be repeated indefinitely until no further change in the estimates takes place.

In an analysis of data set 3 of Table 8.13 the final estimates are practically identical to those obtained by OLS estimation after deleting the outlying observation (see Table 8.15) but with more realistic data this is unlikely to be the case. A similar iterative analysis of data set 1 of Table 8.13 is shown in Table 8.16. The method is called iterative weighted least squares estimation. Note that the weights could have been given a value proportional to the absolute value of any function of the residuals. The ones used in the present analyses are not necessarily the best or the most useful. They were introduced merely to give the student a feel for the method. Other approaches to weighting, and a much more detailed discussion of robust estimation procedures, are given in Mosteller and Tukey (1977). Suitable algorithms for an iterative weighted analysis are available within GLIM.

Table 8.15 Iterative weighted least squares estimation for data set 3 of Table 8.13

Cycle	Estimate of β_0	Estimate of β_1
1 (OLS)	3·00	0·50
2	3·75	0·39
3	3·95	0·35
4	3·99	0·35
5	4·00	0·35

Table 8.16 Iterative weighted least squares estimation for data set 1 of Table 8.13

Cycle	Estimate of β_0	Estimate of β_1
1 (OLS)	3·00	0·50
2	3·06	0·49
3	3·07	0·49
4	3·08	0·49
5	3·09	0·49

8.10 Summary

Regression is one of the most commonly used statistical techniques in the social and behavioural sciences. However, the technique is often misused. The models described in this chapter are not usually appropriate for the analysis of variation in nominal (categorical) or ordinal response variables, but because there are several easily used computer packages that include methods of fitting them, they are often used for data of this type. Social scientists are also prone to investigate the joint effects of too many explanatory variables. Forward section or backward elimination algorithms (Section 8.5) should not be used to replace carefully thoughtout and disciplined research designs. A serious weakness of multiple regression and, indeed, of other modelling techniques, is the possible unreliability of parameter estimates. If one carries out identical analyses on two or more sets of independent data the results are, at times, inconsistent. Methods of tackling this problem are beyond the scope of this text, and the interested reader is referred to Mosteller and Tukey (1977) for a discussion of cross-validation and the use of techniques such as the 'jackknife'.

In this chapter we have only considered the analysis of data for which there is a single response variable. However, the models are easily extended to deal with multivariate responses, leading to methods such as canonical correlation analysis and multivariate analysis of variance (MANOVA). Some of these methods are described in Chapter 11.

Exercises

8.1 From the data given in Table 8.4 fit the following three models, using ordinary least squares estimation:

(a) Visits $= \beta_0 + \beta_1$Age
(b) Visits $= \beta_0 + \beta_2$Weight
(c) Visits $= \beta_0 + \beta_1$Age $+ \beta_2$Weight.
In each case estimate the appropriate parameters (with their standard errors) and note how their value depends on the presence of the other terms in the models.

8.2 Construct appropriate F-statistics to test the fit of the three models in Exercise 8.1. From the regression and residual sums of squares given in Table 8.5, assess the statistical significance of the effects of age and weight on visits to a physician. Should an interaction term be added?

8.3 Use weighted least squares estimation to determine the effect of age and sex on the means of number of symptoms given in Table 8.4.

8.4 Use iterative weighted least squares estimation to fit model (c) of Exercise 8.1. Discuss how you would choose appropriate iterative weights and compare the results of a few of your alternative choices.

8.5 In Table 9.3 is given the proportions of men and women in five age groups who admit to being prescribed psychotropic drugs in a given fortnight. It can be shown that if the proportions, say p_i, are transformed to $\log \frac{p_i}{1-p_i}$ the resulting measure has a variance of

$$\frac{1}{n_i p_i(1 - p_i)}$$

where n_i is the total number in the appropriate category (see Cox and Snell, 1989). Use weighted least squares to estimate the effects of age and sex on psychotropic drug prescription, and compare the results with those given in Chapter 9.

8.6 Consider Table 8.17. Using weighted least squares estimation, fit a regression model to the effects of sex and age on the mean number of symptoms.

8.7 Re-analyse the data in Table 8.10 using forward selection procedure rather than backward elimination (see Table 8.12).

8.8 Consider Table 8.10. Plot TOTCR against each of the either dependent variables. Suggest ways in which the analysis of these data in Table 8.12 might be improved.

8.9 Repeat Exercise 8.7 after first transforming TOTCR by taking its square root.

Table 8.17   Relationship between sex, age and number of symptoms in a sample from West London (Tarnopolsky and Morton-Williams, 1980)

Men: number of symptoms	Age group				
	1	2	3	4	5
0	121	147	143	49	21
1	117	106	126	57	12
2	117	78	104	43	16
3	79	73	86	43	10
4	70	51	70	34	14
5	72	44	60	27	10
6	32	39	47	26	5
7	32	28	32	12	6
8	17	18	24	10	6
9	9	18	15	8	6
10	9	14	18	7	1
11	10	3	18	7	5
12	9	2	11	3	2
13	3	2	1	4	1
14	3	1	9	0	0
15	2	1	6	3	2
16	0	2	3	1	0
17	1	0	1	3	0
18	1	0	2	0	0
19	0	0	0	0	1
20	0	1	0	1	0
21	0	0	0	0	0

Women: number of symptoms	Age group				
	1	2	3	4	5
0	105	115	101	41	22
1	89	86	111	43	25
2	84	93	111	41	27
3	93	95	98	43	29
4	77	90	110	45	20
5	76	59	67	45	19
6	60	62	90	37	17
7	47	55	65	25	16
8	38	31	54	30	21
9	25	24	61	15	11
10	36	22	32	18	13
11	13	19	30	13	5
12	12	8	25	10	7
13	13	14	19	13	7
14	7	6	15	6	2
15	2	6	12	1	3
16	1	1	9	2	1
17	3	0	3	2	0
18	2	3	1	1	0
19	1	0	0	1	1
20	0	0	1	0	1
21	0	0	1	1	0

9

Linear Models for Categorical Data

9.1 Introduction

The techniques described in the last chapter involved linear models for continuous response variables with normally distributed random disturbance terms. In this chapter we shall consider linear models suitable for categorical and, in particular, for binary and ordinal response variables. These will involve the use of the logarithmic and logistic link functions together with Poisson and binomial probability distributions, respectively.

In order to examine the relationships between categorical variables the data are usually arranged in the form of a contingency table. Table 9.1, for example, is a simple two-way contingency table arising from the cross-classification of a sample of 5883 West Londoners with respect to their age group and sex. Tables 9.3 and 9.5 are examples of contingency tables with more than two classifying factors (or dimensions). We shall assume here that readers are familiar with the simple analysis of such tables by means of the Pearson chi-square statistic, and with the more common measures of association such as the odds-ratio. (Details of both topics are available in Everitt 1977a.)

9.2 Maximum Likelihood Estimation for Log-Linear and Logistic Models

To obtain the maximum likelihood estimates for a particular model we need to consider the sampling distributions of the observations. Three commonly encountered sampling plans lead to different distributions.

(1) *Independent Poisson sampling*: with the restrictions on the sample size, N, each count y in a contingency table has an independent Poisson distribution with mean m_i, and the likelihood is given by

$$L = \prod_i \frac{m_i^{y_i} \exp(-m_i)}{y_i!} \qquad (9.1)$$

Such a situation arises if observations are made over a period of time with no *a priori* knowledge of the total number of observations.

Table 9.1   A sample of West Londoners cross-classified by age and sex (taken from Murray *et al.*, 1981)

Age	16–29	30–44	45–64	65–74	Over 75	Total
Male	704	628	775	338	118	2563
Female	784	789	1016	434	247	3270
Total	1488	1417	1791	772	365	5833

As we have seen in Chapter 7, the expected count m_i is related to the explanatory variables by a relationship of the form

$$m_i = \exp(\sum_j \beta_j x_{ij}) \qquad (9.2)$$

where the x_{ij} are appropriate dummy variables, and the β_j are the corresponding parameters or regression coefficients that are to be estimated. Substitution of equation (9.2) into (9.1) followed by the usual maximum likelihood process leads to the required estimates, although details are outside the scope of this text. Interested readers are referred to Agresti (1990) and Bishop *et al.* (1975).

(2)   *Simple multinomial sampling*: sometimes the total sample size, N, is fixed by design. This restriction imposed on a series of independent Poisson distributions leads to a multinomial distribution and a likelihood function

$$L = \frac{N!}{\prod_i y_i!} \prod_i (\frac{m_i}{N})^{y_i} \qquad (9.3)$$

Again equation (9.2) would be substituted into this function to get the maximum likelihood estimates of the β_j.

(3)   *Product multinomial sampling*: although in many observational studies only a single sample will be examined, in many situations (including experiments) it is more usual to have several groups, with the total number of individuals in each group determined by the sampling plan. If the total count in the jth group is fixed at N_j the likelihood is then given by

$$L = \prod_j \frac{N_j!}{\prod_i y_{ij}!} \prod_i (\frac{m_{ij}}{N_j})^{y_{ij}} \qquad (9.4)$$

Where m_{ij} and y_{ij} are the expected counts and observed counts in the ith cell of the jth stratum, respectively. Here, the likelihood is seen to be the product of several multinomial density functions of the form of equation (9.3). Again, the substitution of an appropriate form of (9.2) for the m_{ij} followed by differentiation, and so on, will lead to estimates of the β_j.

In practice, it is easier to consider the logarithm of the likelihood, L. Ignoring terms not involving the parameters of the model, equations (9.1), (9.3) and (9.4) yield the following log-likelihood equations respectively:

$$L = \sum_i y_i \ln(m_i) - \sum_i m_i, \qquad (9.5)$$

$$L = \sum_i y_i \ln(m_i), \text{ and} \tag{9.6}$$

$$L = \sum_{i,j} y_{ij} \ln(m_{ij}) \tag{9.7}$$

On the assumption that the sum of the estimates of the m_i in (9.5) will be constrained to be N, these three expressions have an identical form. The likelihoods for the three different sampling distributions are said to have an identical *kernel*. When fitting log-linear models using a program such as GLIM the Poisson error distribution is specified in all three cases. The only stage where the sampling distribution has to be recognised is in the specification of the linear predictors within equation (9.2).

(4) *Likelihood function for logistic models*: here we assume that the ith cell of the contingency table contains counts for two types of individual, a total of r_i being positive for some criterion and $n_i - r_i$ negative. In the population, we assume that the probability of an individual in the ith cell being scored as positive is p_i and accordingly the probability of being negative is $1 - p_i$. The likelihood for this situation is simply the product of the binomial distributions appropriate for each cell:

$$L = \prod_i \frac{n_i!}{r_i!(n_i - r_i)!} p_i^{r_i}(1 - p_i)^{n_i - r_i} \tag{9.8}$$

with, as we have seen in Chapter 7, the probabilities p_i being related to the explanatory variables as follows:

$$p_i = \frac{\exp(\sum_j \beta_j x_{ij})}{1 + \exp(\sum_j \beta_j x_{ij})} \tag{9.9}$$

Equation (9.8) is of course, a special case of the likelihood for the product multinomial sampling scheme given in (9.4).

9.3 Goodness-of-Fit Measures and the Comparison of Alternative Models

Maximum likelihood estimation for log-linear and linear-logistic models leads, in addition to estimates for the parameters β_j, to a set of estimated or predicted cell counts under the model currently being entertained. The goodness of fit of this model is then assessed by comparison of these to the observed counts by means of either of two statistics, namely

$$X^2 = \sum_{\text{all cells}} \frac{(\text{observed} - \text{expected})^2}{\text{expected}} \tag{9.10}$$

or

$$G^2 = 2 \sum (\text{observed}) \ln\left(\frac{\text{observed}}{\text{expected}}\right) \tag{9.11}$$

where the summation in both cases is over all cells in the table. X^2 is the familiar Pearson chi-square statistic; G^2 is a log-likelihood ratio criterion or *deviance*. If the current model is adequate and the sample size is relatively large, both statistics are approximately distributed as χ^2 with degrees of freedom (d.f.) given by

$$d.f. = \text{No. of cells in table}$$
$$- \text{No. of independent parameters estimated} \qquad (9.12)$$

In general we shall use the less familiar deviance criterion, G^2. This statistic will also be used for the comparison of models. Thus, if $G^2(a)$ is the value for a model with n_a d.f. and $G^2(b)$ is the value for a more complex model (derived from the first by the addition of extra parameter) with n_b d.f., then the quantity

$$G^2(b|a) = G^2(a) - G^2(b) \qquad (9.13)$$

can be used to assess whether the second model has significantly improved the fit. This change in deviance can be compared with χ^2 with $n_a - n_b$ d.f. This procedure is analogous to the use of regression sum-of-squares and F-tests in the previous chapter, and the forward and backward selection procedures described here may also be used in attempting to find an adequate model. Examples of this will be given below.

9.4   Some Numerical Examples

Let us begin by examining the application of log-linear and logistic models to the simple 2×2 table shown in Table 9.2(a). Here we are interested in the possible association between sex and psychotropic drug prescriptions. The sample of West Londoners described by Table 9.1 were simply asked whether they had received a prescription for a psychotropic drug within the fortnight prior to interview (for further details see Murray *et al*, 1981). The odds-ratio estimated from Table 9.2(a) is 2.07 and the corresponding natural logarithm is 0.73 (with estimated standard error 0.09). The odds on women being prescribed psychotropic drugs is about twice that for men. Now consider fitting the log-linear model

$$\ln(m_{ij}) = \beta_0 + \text{Sex}(i) + \text{Drug}(j) + \text{Sex}(i).\text{Drug}(j) \qquad (9.14)$$

where $i = 1, 2$ (men, women, respectively) and $j = 1, 2$ (prescribed drugs, not prescribed drugs, respectively). $\text{Sex}(i)$, $\text{Drug}(j)$ and $\text{Sex}(i).\text{Drug}(j)$ are parameters representing main effects of sex and drug prescription and the two-way interaction between them. (This model could, of course, be rewritten in terms of suitable dummy variables as indicated in the two previous chapters.) As written, the model has nine parameters, but since the table has only four cells, some of these parameters will not have unique estimates. The model could be re-parameterised in terms of estimate contrasts or constraints on the parameter introduced. As mentioned in the previous two chapters, GLIM uses the latter approach and sets a number of the parameters to zero (these are $\text{Sex}(1)$, $\text{Drug}(1)$, $\text{Sex}(1).\text{Drug}(1)$, $\text{Sex}(2).\text{Drug}(1)$ and $\text{Sex}(1).\text{Drug}(2)$). Estimates of the unconstrained parameters are given in Table 9.2(b). Since the estimated expected values under this model are equal to the observed values, G^2 takes the value zero (with 0 d.f.). A model with the same number of free parameters as the number of cells is known as the saturated model. If we now fit a model which sets all of the interaction terms to zero, that is

$$\ln(m_{ij}) = \beta_0 + \text{Sex}(i) + \text{Drug}(j) \qquad (9.15)$$

Table 9.2 Cross-classification of a sample of West Londoners according to sex and psychotropic drug prescription (from Murray *et al.*, 1981)

(a) Observed counts

		Sex	
		Men(1)	Women (2)
Drug	Not prescribed (1)	2378	2817
	Prescribed (2)	185	453

(b) Log-linear model: $\ln(m_{ij}) = \beta_0 + Sex(i) + Drug(j) + Sex(i).Drug(j)$

Parameter	Estimate	Standard error
β_0	7·77	0·02
Sex(2)	0·17	0·03
Drug(2)	−2·55	0·08
Sex(2).Drug(2)	0·73	0·09

(c) Log-linear model: $\ln(m_{ij}) = \beta_0 + Sex(i) + Drug(j)$

Parameter	Estimate	Standard error
β_0	7·73	0·02
Sex(2)	0·24	0·03
Drug(2)	−2·10	0·04
$G^2 = 67·38$ with 1 d.f.		

(d) Linear-logistic model: $\ln\left(\dfrac{m_{i2}}{m_{i1}}\right) = \beta_0 + Sex(i)$

Parameter	Estimate	Standard error
β_0	−2·55	0·08
Sex(2)	0·73	0·09

(e) Linear-logistic model: $\ln\left(\dfrac{m_{i2}}{m_{i1}}\right) = \beta_0$

Parameter	Estimate	Standard error
β_0	−2·10	0·04
$G^2 = 67·38$ with 1 d.f.		

we obtain the estimates as shown in Table 9.2 (c) and a G^2 value of 67.38 with a single degree of freedom. This G^2 provides a test for the significance of the two-way interaction in model (9.14). Note that the estimate of the two-way interaction given in Table 9.2(b) is the logarithm of the odds-ratio given above. The square of the ratio of this estimate to its standard error is 62.85 and this too can also be compared with a chi-square variate with a single degree of freedom. In the context of the present analysis the parameters β_0, Sex(i) and Drug(j) are of no interest — they are often referred to as *nuisance parameters*.

If we now reformulate our interest in these data as that of investigating the effect of sex on drug prescription, we might consider fitting a logistic model of the form

$$\ln \frac{p_i}{1 - p_i} = \beta_0 + Sex(i) \qquad (9.16)$$

(The parameters β_0 and Sex(i) are not the same as those in (9.14) and (9.15).) Here p_i is the probability that an individual of sex i ($i = 1$ for men and 2 for women) is prescribed psychotropic drugs. Fitting this model using GLIM results in the parameter estimates given in Table 9.2(d). Again G^2 is zero since this model fits the data perfectly. If we fit a model assuming that sex has no effect, that is

$$\ln \frac{p_i}{1 - p_i} = \beta_0 \qquad (9.17)$$

then we obtain the results shown in Table 9.2(c). G^2 now has the value 67.38 and consequently we are led to exactly the same conclusion as in the previous analysis using log-linear models. The required log(odds-ratio) is now provided by the estimate of Sex(2) and is the same as that for Sex(2).Drug(2) in the log-linear models. The equivalence of some of the parameter estimates in the two types of model is easily explained if we return to the examination of log-linear models for people not prescribed drugs, that is

$$\ln(m_{i1}) = \beta_0 + \text{Sex}(i) \qquad (9.18)$$

and for those prescribed drugs

$$\ln(m_{i2}) = \beta_0 + \text{Sex}(i) + \text{Drug}(2) + \text{Sex}(i).\text{Drug}(2) \qquad (9.19)$$

(remembering that some parameters are *a priori* set to zero).
  Subtracting equation (9.18) from (9.19) gives

$$\ln(\frac{m_{i2}}{m_{i1}}) = \text{Drug}(2) + \text{Sex}(i).\text{Drug}(2) \qquad (9.20)$$

Sex(i) in the logistic model is equivalent to Sex(i).Drug(2) in the corresponding log-linear model. Similarly β_0 of the logistic model is equivalent to *Drug*(2) in the log-linear model (here estimated by $- 2.55$).

As an example of a three-dimensional table we will consider the dates shown in Table 9.3. This gives the number of people who admit to have been prescribed psychotropic drugs in the fortnight prior to interview as a function of both age and sex. The two-way table in Table 9.2 is, in fact, obtained by adding over the age groups of Table 9.3. Suppose that we are still primarily interested in estimating the log-odds ratio measuring the influence of sex on drug prescriptions. Now, however, we are in a position to test for any confounding effects of age. Typically, epidemiologists are concerned with testing for possible heterogeneity of the log-odds ratio across the five age groups and, if this is not significant, in the estimation of the log-odds ratio that is common to all age-groups (see, for example, Armitage and Berry, 1987, pp. 458–60). We first approach this problem though the use of log-linear models. Consider the effects of sex (Sex(i), $i = 1, 2$), age group (Agegp(j), $j = 1$ *to* 5) and drug prescription (Drug(k), $k = 1, 2$). We start by fitting

$$\ln(m_{ijk}) = \beta_0 + \text{Sex}(i) + \text{Agegp}(j) + \text{Drug}(k) + \text{Sex}(i).\text{Agegp}(j)$$
$$+ \text{Sex}(i).\text{Drug}(k) + \text{Agegp}(j).\text{Drug}(k) \qquad (9.21)$$

The deviance, G^2, is 2.30 with 4 degrees of freedom. This provides a test for the goodness of fit of the model and, equivalently a test for the three-way interaction, Sex(i).Agegp(j).Drug(k). The latter interaction is a measure of heterogeneity of the required two-way interaction Sex(i).Drug(k). Removal of Sex(i).Drug(k) from (9.21) would provide a test of significance for the

Table 9.3   Patterns of psychotropic drug consumption in a sample from West London (taken from Dunn, 1981)

Sex	Age group	Mean age	No. taking drugs	No. not on drugs	Odds	log (odds)
Male	16–29	23·2	21	683	0·03	−3·5
Male	30–44	36·5	32	596	0·05	−2·9
Male	45–64	54·3	70	705	0·10	−2·3
Male	65–74	69·2	43	295	0·15	−1·9
Male	over 74	79·5	19	99	0·19	−1·7
Female	16–29	23·2	46	738	0·06	−2·8
Female	30–44	36·5	89	700	0·13	−2·1
Female	45–64	54·3	169	847	0·20	−1·6
Female	65–74	69·2	98	336	0·29	−1·2
Female	over 74	79·5	51	196	0·26	−1·3

common log-odds ratio measuring the influence of sex on psychotropic drug prescription. Note that this is a conditional test. It is a test of the significance of the influence of sex on drug prescription after allowing for the association between sex and age group and for the influence of age group on drug prescription. The estimate of Sex(2).Drug(2) is 0.69 (standard error 0.09) equivalent to a common odds-ratio of 2.0.

Now consider logistic models for the proportion being prescribed drugs as a function of age group and sex. The logistic model equivalent to (9.21) is

$$\ln(\frac{m_{ij2}}{m_{ij1}}) = \beta_0 + \text{Sex}(i) + \text{Agegp}(j) \tag{9.22}$$

The goodness of fit of this model provides a test of heterogeneity for the required odds-ratio in terms of the significance of the two-way interaction Sex(*i*).Agegp(*j*). The required common estimate is provided by the exponent of Sex(2). It is identical to that provided by fitting the log-linear model in (9.21).

So far, in both (9.21) and (9.22), we have considered the effect of age group as if it were a nominal measurement. Now consider models which take account of the quantitative nature of the age measurements. From the last two columns of Table 9.3 we can calculate an observed log-odds and then plot this statistic as a function of sex and mean age of the corresponding group. This plot is shown in Figure 9.1. The influence of age and sex has been represented here by two straight and parallel lines. The algebraic description of the two parallel lines is given by

$$\ln(\text{odds}) = \beta_0 + \beta_1 \text{Sex} + \beta_2 \text{Age} \tag{9.23}$$

where Sex is now a binary dummy variable (taking the value of 0 for men and 1 for women) and Age is the mean age of the required group. The results of fitting the model in (9.23) are given in Table 9.4(a). The value of G^2 indicates that this model appears to fit the data quite well. Visual examination of the data in Figure 9.1, however, indicates that the model might be improved by allowing for some curvature. This is also illustrated by examination of the

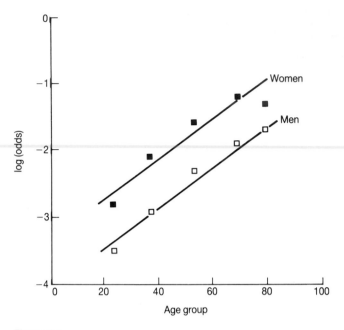

Figure 9.1

column labelled as 'residual' in Table 9.4(a). These residuals (r) are defined by

$$r = \frac{y - n\hat{p}}{\sqrt{n\hat{p}(1 - \hat{p})}} \tag{9.24}$$

where y is the observed number of people being prescribed drugs in a particular age group and sex combination, and $n\hat{p}$ is the corresponding fitted value. The total number of people in the group is given by n. These are standardised residuals equivalent to those described in Section 8.7. They are often referred to as Pearson residuals, the sum of the squares of these residuals giving the Pearson chi-square statistic. Other forms of residual and further diagnostic checks are described in Agresti (1990). The corresponding Pearson residual for counts in a contingency table modelled by a log-linear model are

$$r = \frac{n - m}{\sqrt{m}} \tag{9.25}$$

where n is the observed cell count and m the corresponding fitted value.
Returning to the residuals given in Table 9.4(a) we see that they tend to negative at both ends of the age range and positive in the centre. A better fitting model is provided by

$$\ln(\text{odds}) = \beta_0 + \beta_1 \text{Sex} + \beta_2 \text{Age} + \beta_3 \text{Age}^2 \tag{9.26}$$

Here the parameter β_3 is a measure of the quadratic effect of ageing on the logit of drug prescriptions. Parameter estimates for this model are given in

Table 9.4 Models for psychotropic drug prescription (Table 9.3)

(a) Model: $\ln(odds) = \beta_0 + \beta_1 Sex + \beta_2 Age$

Parameter	Estimate	Standard error
β_0	$-3 \cdot 90$	$0 \cdot 15$
β_1	$0 \cdot 69$	$0 \cdot 09$
β_2	$0 \cdot 03$	$0 \cdot 002$

$G^2 = 11 \cdot 56$ with 7 d.f.

Sex	Age	Observed	Total	Fitted	Residual	
0	23·2	21	704	26·17		−1·03
0	36·5	32	628	33·32		−0·24
0	54·3	70	775	65·44		0·59
0	69·2	43	338	41·50		0·25
0	79·5	19	118	18·57		0·11
1	23·2	46	784	55·79		−1·36
1	36·5	89	789	78·94		1·19
1	54·3	169	1016	157·16		1·03
1	69·2	98	434	94·33		0·43
1	79·5	51	247	66·78		−2·26

(b) Model: $\ln(odds) = \beta_0 + \beta_1 Sex + \beta_2 Age + \beta_3 Age^2$

Parameter	Estimate	Standard error
β_0	$-4 \cdot 85$	$0 \cdot 37$
β_1	$0 \cdot 69$	$0 \cdot 09$
β_2	$0 \cdot 07$	$0 \cdot 015$
β_3	$-0 \cdot 0004$	$0 \cdot 00014$

$G^2 - 3 \cdot 33$ with 6 d.f.

Table 9.4(b). The reduction in G^2 in going from model (9.23) to (9.26) provides a test of significance for β_3 (change in G^2 is 8.23 with 1 d.f.).

9.5 Model Selection Procedures

Consider the data shown in Table 9.5 taken from Ries and Smith (1963). These data arise from a study in which a sample of 1008 people was asked to compare two detergents, brand M and brand X. In addition to stating their brand preference, the sample members provided information on previous use of brand M (yes or no), the degree of softness of the water they used (soft, medium, or hard), and the temperature of the water (high, low). The main topic of interest here is the effect of the variables — previous use, degree of softness and temperature — on brand preference (M as opposed to X). This would suggest the use of a linear-logistic model. Initially, however, we will use log-linear models to explore associations between the four variables (ignoring the distinction between explanatory and response variables) through the use of forward and backward selection procedures analogous to those described in Section 8.5.

Table 9.5 Cross-classification of a sample of 1008 consumers in a blind trial according to (1) water softness, (2) previous use of detergent brand M, (3) water temperature, and (4) preference for brand M (Ries and Smith, 1963)

Water softness	Brand preference	*Previous user of M (Prev = 1)*		*Not previous user of M (Prev = 2)*	
		High temp. (Temp = 2)	*Low temp. (Temp = 1)*	*High temp.*	*Low temp.*
Soft (Soft = 1)	X (Brand = 1)	19	57	29	63
	M (Brand = 2)	29	49	27	53
Medium (Soft = 2)	X	23	47	33	66
	M	47	55	23	50
Hard (Soft = 3)	X	24	37	42	68
	M	43	52	30	42

We start by choosing an arbitrary significance level of, say, 0.05 (but see below) and use it to assess the fit of the following three log-linear models

(A) Brand + Prev + Soft + Temp   (9.27)

(B) (A) + Brand.Prev + Brand.Soft + Brand.Temp
+ Prev.Soft + Prev.Temp + Soft.Temp   (9.28)

(C) (B) + Brand.Prev.Soft + Brand.Prev.Temp + Brand.Soft.Temp
+ Prev.Soft.Temp   (9.29)

where the variable names have a meaning that is obvious within the context of this example, and the subscripts indicating variable categories have been dropped to simplify the nomenclature. Terms such as Brand.Prev are used for first-order interaction parameters, and Brand.Prev.Soft for second-order interactions. Model (A) contains only main effects, model (B) has main effects and first-order interactions, and model (C) contains main effects and both first- and second-order interactions. The results of fitting each of these models is shown at the top of Table 9.6. Model (A) does not describe the data adequately, but (B) and (C) do. Clearly we need a model more complex than (A) but possibly simpler than (B). Consequently, we can proceed by either forward selection of interaction terms to add to (A), or backwards elimination of terms from (B) to search for the 'best' model.

In forward selection we add each two-factor interaction in turn and choose the one for which the change in the conditional goodness-of-fit criterion of equation (9.13) is most significant from Table 9.6, namely Brand.Prev. Next, having included Brand.Prev in the model, we add the remaining two-factor interacting terms in turn, and again choose the one for which the change in the conditional goodness-of-fit statistic is most significant. This leads to the addition of the term Brand.Temp. Repeating this procedure we find that we must add Soft.Temp but that addition of further terms causes no significant improvement in the fit. Hence we are led to the following model as the simplest that adequately describes the data:

$$(A) + \text{Brand.Prev} + \text{Brand.Temp} + \text{Soft.Temp} \qquad (9.30)$$

As an optional step, after the addition of each of the two-factor interaction terms, we might have considered deleting any of the terms currently included that no longer contributed significantly to the fit. However, in this example no terms could be deleted.

Backwards elimination is a very similar procedure, but here we start with model (B) and delete each two-factor interaction in turn. First, the least significant term is deleted, and so on. At each step we add back two-factor interaction terms that now significantly improve the fit of the current model. The results of applying this procedure to the Ries–Smith data are also shown in Table 9.6. The 'best' model is

$$(B) - \text{Soft.Brand} - \text{Soft.Prev} - \text{Prev.Temp} \qquad (9.31)$$

In this case, it is the same model as that found by forward selection, but this is not always so. The estimated parameters of the model are given in Table 9.7.

Returning to the idea of fitting a linear-logistic model to the proportions of people choosing brand M, appropriate starting points would be

(D)	Const	(9.32)
(E)	Prev + Soft + Temp	(9.33)
(F)	(E) + Prev.Soft + Prev.Temp + Soft.Temp	(9.34)

where models (D), (E) and (F) are superficially similar to (A), (B) and (C), respectively, but not quite. Model (E), for example is equivalent to the following log-linear model:

$$(A) + \text{Prev.Soft} + \text{Prev.Temp} + \text{Soft.Temp} + \text{Prev.Soft.Temp}$$
$$+ \text{Prev.Brand} + \text{Soft.Brand} + \text{Temp.Brand} \qquad (9.35)$$

In the terminology of GLIM, (D), (E) and (F) are, respectively, equivalent to

$$\text{Prev} * \text{Soft} * \text{Temp} + \text{Brand} \qquad (9.36)$$

$$\text{Prev} * \text{Soft} * \text{Temp} + \text{Brand} + \text{Prev.Brand} + \text{Soft.Brand} + \text{Temp.Brand}$$
$$(9.37)$$

$$\text{Prev} * \text{Soft} * \text{Temp} + \text{Brand} + \text{Prev.Brand} + \text{Soft.Brand} + \text{Temp.Brand}$$
$$+ \text{Prev.Soft.Brand} + \text{Prev.Temp.Brand} + \text{Soft.Temp.Brand} \qquad (9.38)$$

Fitting a series of linear-logistic model is equivalent to fitting the corresponding log-linear models in which the models always contain all possible interactions between the explanatory variables (here represented by Prev*Soft*Temp). The results for the Ries–Smith data are given in Table 9.7.

Details of other model-selection strategies are given by Agresti (1990, Chapter 7). One important problem for the above selection procedures, and for several others, is that caused by the repeated use of significance tests on the same set of data. Simultaneous test procedures to allow for this are discussed by Aitkin (1979, 1980) and by Whittaker and Aitkin (1978). Regardless of the model selection strategy used, one must not ignore the sampling design.

Table 9.6   Log-linear models: analysis of G^2 for Ries–Smith data

Model	G^2	d.f.
(A)	42·93	18
(B)	9·85	9
(C)	0·74	2

1 Forward selection

	G^2	d.f.
(A) + Soft × Brand	42·53	16
(A) + Soft × Prev	41·85	16
(A) + Soft × Temp	36·83	16
(A) + Brand × Prev	22·35	17
(A) + Brand × Temp	38·57	17
(A) + Prev × Temp	41·68	17
(A) + Brand × Prev + Soft × Brand	21·95	15
(A) + Brand × Prev + Soft × Prev	21·27	15
(A) + Brand × Prev + Soft × Temp	16·25	15
(A) + Brand × Prev + Brand × Temp	17·99	16
(A) + Brand × Prev + Prev × Temp	21·09	16
(A) + Brand × Prev + Brand × Temp + Soft × Brand	17·59	14
(A) + Brand × Prev + Brand × Temp + Soft × Prev	16·91	14
(A) + Brand × Prev + Brand × Temp + Soft × Temp	11·89	14
(A) + Brand × Prev + Brand × Temp + Prev × Temp	17·29	15

$\vdots$

etc.

2 Backward elimination

	G^2	d.f.
(B) − Soft × Brand	10·06	11
(B) − Soft × Prev	10·85	11
(B) − Soft × Temp	15·94	11
(B) − Brand × Prev	29·74	10
(B) − Brand × Temp	13·58	10
(B) − Prev × Temp	10·59	10
(B) − Soft × Brand − Soft × Prev	11·19	13
(B) − Soft × Brand − Soft × Temp	16·22	13
(B) − Soft × Brand − Brand × Prev	30·08	12
(B) − Soft × Brand − Brand × Temp	13·86	12
(B) − Soft × Brand − Prev × Temp	10·80	12
(B) − Soft × Brand − Soft × Prev − Soft × Temp	17·29	15
(B) − Soft × Brand − Soft × Prev − Brand × Prev	31·21	14
(B) − Soft × Brand − Soft × Prev − Brand × Temp	14·99	14
(B) − Soft × Brand − Soft × Prev − Prev × Temp	11·89	14

$\vdots$

etc.

How this is done for linear-logistic models should be clear because of their similar structure to the more familiar ANOVA and ANCOVA models. In log-linear models one should always include interactions which are equivalent to fixing the corresponding table margins that are fixed by the sampling design. Consider Table 9.8 for example. This is a four-way table: Age by Sex by GHQ by Drug prescription (remembering that, in this example, Age is a categorical variable). Suppose, for the purposes of the present argument, that the numbers

Table 9.7 Parameter estimates for Ries–-
Smith data

---

*(i) Log-linear model: Brand + Prev + Soft +
Temp + Brand × Prev + Brand × Temp + Soft ×
Temp*

Parameter	Estimate	Standard error
β_0	3·87	0·09
Soft (2)	−0·02	0·10
Soft (3)	−0·11	0·10
Brand (2)	0·18	0·10
Prev (2)	0·37	0·09
Temp (2)	−0·90	0·14
BP (22)	−0·56	0·13
ST (22)	0·21	0·16
ST (32)	0·40	0·16
BT (22)	0·27	0·13
	$G^2 = 11·89$ with 14 d.f.	

(ii) Linear-logistic model: Prev + Temp

Parameter	Estimate	Standard error
β_0	0·19	0·10
Prev (2)	−0·57	0·13
Temp (2)	0·26	0·13
	$G^2 = 8·44$ with 9 d.f.	

(iii) Linear-logistic model: Prev

Parameter	Estimate	Standard error
β_0	+0·29	0·09
Prev (2)	−0·58	0·13
	$G^2 = 12·24$ with 10 d.f.	

---

of each age-group for each of the two sexes had been fixed by the sampling
design. This would imply that all fitted log-linear models must include the
Age.Sex interaction. The simplest model to be considered would be

$$\text{Age} * \text{Sex} + \text{GHQ} + \text{Drug} \qquad (9.39)$$

where Age*Sex is equivalent to Age + Sex + Age.Sex, and the meanings of
names of the model terms should be clear from the context. The data in
Table 9.8 can, in fact, be regarded as arising for a single unstratified random
sample. In modelling these data, however, it would still appear to be sensible
to acknowledge the different status of each of the four variables. Age and Sex
could still be regarded as explanatory variables and either or both of GHQ and
Drug might be regarded as responses. If drug prescription is regarded as the
response of interest, and the rest are considered to be explanatory variables,
then the appropriate starting point is

$$\text{Age} * \text{Sex} * \text{GHQ} + \text{Drug} \qquad (9.40)$$

If, however, psychiatric status is assumed to be influenced by age, sex and drug
taking, the following may be more appropriate:

$$\text{Age} * \text{Sex} * \text{Drug} + \text{GHQ} \qquad (9.41)$$

Table 9.8 Patterns of psychotropic drug consumption in a sample from West London (from Murray *et al.*, 1981)

Sex	Mean age	Probable psychiatric case*	No. taking drugs	Total
Male	23·2	No	9	531
Male	36·5	No	16	500
Male	54·3	No	38	644
Male	69·2	No	26	275
Male	79·5	No	9	90
Male	23·2	Yes	12	171
Male	36·5	Yes	16	125
Male	54·3	Yes	31	121
Male	69·2	Yes	16	56
Male	79·5	Yes	10	26
Female	23·2	No	12	568
Female	36·5	No	42	596
Female	54·3	No	96	765
Female	69·2	No	52	327
Female	79·5	No	30	179
Female	23·2	Yes	33	210
Female	36·5	Yes	47	189
Female	54·3	Yes	71	242
Female	69·2	Yes	45	98
Female	79·2	Yes	21	60

* High or low score on completion of the General Health Questionnaire (D. P. Goldberg, 1972)

Perhaps a more sensible approach would be to model the joint distribution of Drug and GHQ by Age and Sex. Here the starting point would be

$$\text{Age} * \text{Sex} + \text{GHQ} * \text{Drug} \tag{9.42}$$

Here an interesting question to ask is whether the association between psychiatric status and psychotropic drug prescription (the GHQ.Drug interaction) is influenced by age and sex. Aitkin *et al.* (1989) refer to models such as that in equation (9.42) as models for structured multinomial responses and the reader is referred to this text for further details. Finally, consider a causal hypothesis in which age and sex jointly influence psychiatric status and the combination of these three then predicts psychotropic drug prescription. This would imply that one might first ignore drug prescription and model the influence of Age and Sex on GHQ. The starting point would now be

$$\text{Age} * \text{Sex} + \text{GHQ} \tag{9.43}$$

The starting point for the second part of this causal hypothesis would be (9.40). Details of model-fitting strategies for this situation are discussed in Agresti (1990, Chapter 7) and Fienberg (1980, Chapter 7).

9.6   Linear Models for Ordinal Response Variables

Consider the data shown in Table 9.9, which comprises counts of men and women in the West London sample, first considered in Table 9.1, but now cross-classified with respect to the same five age groups and their perceptions

Table 9.9   Effect of sex and age on self-assessment of health (from Murray *et al.*, 1982)

		Health (Men)				
		1	2	3	4	5
	1	271	233	151	43	4
	2	243	185	140	44	6
Age-group:	3	243	235	218	60	14
	4	104	87	94	38	11
	5	21	38	35	17	3
		Health (Women)				
		1	2	3	4	5
	1	246	242	211	70	6
	2	242	237	216	78	9
Age-group:	3	223	280	333	145	24
	4	92	101	147	71	16
	5	43	53	83	42	24

of their own health over the two weeks prior to being interviewed. This assessment is made on an ordinal five-point scale ranging from 'very good' (coded as 1) to 'very poor' (coded as 5). Of interest here is the manner in which the sex and age of the respondent influence self-assessment of health. We start by considering log-linear models for these data. Treating age group and health as qualitative factors each with five levels we fit the minimal model

$$\text{Age} * \text{Sex} + \text{Health} \qquad (9.44)$$

giving a deviance of 281.90 with 36 degrees of freedom. If we now let age group and sex both influence health assessment by fitting

$$\text{Age} * \text{Sex} + \text{Health} + \text{Age.Health} + \text{Sex.Health} \qquad (9.45)$$

we obtain a deviance of 11.49 with 16 degrees of freedom. Although this provides a good fit to the data and suggests that the effects of age and sex are additive (no three-way interaction is needed), it has not provided much of a simplification. There are twenty interaction parameters to be estimated. The ordinal nature of both age group and health status has not been taken into account. One possibility of simplification is to consider only the linear components of Age and Health in the above two-way interactions. This gives an example of a linear-by-linear association model which will be discussed later in this section. Here we consider approaches using linear-logistic models for proportions.

To simplify the problem, first consider the marginal counts, ignoring the sex of the subject (see Table 9.10). Initially, we might think of terms of a model for the dichotomous variable produced by dividing the table between two arbitrary assessments, say, between health status of 2 and 3. Plotting the log(odds) for an assessment of 3 or more against age group leads to Figure 9.2. We can also consider other places in the table at which to make these 'cuts' (there being four in all), and again plot the log(odds) for assessing our own

Table 9.10 Self-assessment of health in a West London Sample: data from Table 9.9 collapsed over sex

(a) Counts

		Health				
		1	2	3	4	5
	1	517	475	362	113	10
	2	485	422	356	122	15
Age	3	466	515	551	205	38
group	4	196	188	241	209	27
	5	64	91	118	59	27

(b) Cumulative odds

		Cut 1		Cut 2		Cut 3		Cut 4	
		No. less	No. above	No. less	No. above	No. less	No. above	No. less	No. above
	1	517	960	992	485	1354	123	1467	10
	2	485	915	907	493	1263	137	1385	15
Age	3	466	1309	981	794	1532	243	1737	38
group	4	196	665	384	477	625	236	834	27
	5	64	295	155	204	273	86	332	27

(c) Conditional odds

		Cut 1		Cut 2		Cut 3		Cut 4 KD5	
		No. less	No. above	No. less	No. above	No. less	No. above	No. less	No. above
	1	517	960	475	485	362	123	13	10
	2	485	915	422	493	356	137	122	15
Age	3	466	1309	515	794	551	243	205	38
group	4	196	665	188	477	241	236	209	27
	5	64	295	91	204	118	86	59	27

health above the cut against age group. Figure 9.3 shows such a plot for a cut between 3 and 4. Table 9.10(b), illustrates the use of all four cuts to generate the required odds for these plots.

The data in Table 9.10(b) could be investigated by fitting a series of linear-logistic models with age group and cut as explanatory variables. For example, a model such as

$$\ln(\text{odds}) = \text{Cut} + \text{Age} + \text{Age.Cut} \qquad (9.46)$$

where Cut and Age are qualitative factors, with subscripts dropped for simplicity, would merely provide an alternative to the saturated log-linear model for the data as presented in Table 9.10(a). However, if in the plots of log(odds) for different cutting points the four lines appeared to be parallel, then an appropriate model would be

$$\ln(\text{odds}) = \text{Cut} + \text{Age} \qquad (9.47)$$

Finally, if one were to postulate that there was no effect of age-group on self-assessment of health, the appropriate model would be

$$\ln(\text{odds}) = \text{Cut} \qquad (9.48)$$

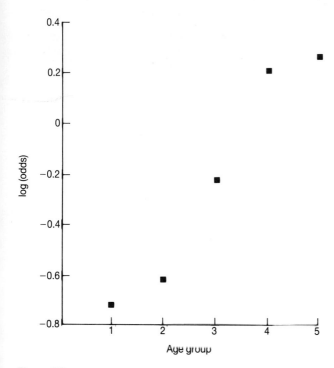

Figure 9.2

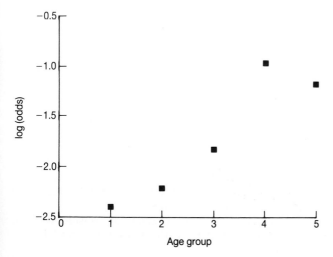

Figure 9.3

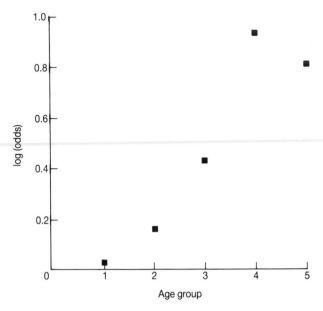

Figure 9.4

This class of models has been described by McCullagh (1980) and we shall refer to them as cumulative adds models.

Returning to Table 9.10(a) we could decide to rearrange and plot the data in a slightly different way. We could, for instance, select those subjects with a self-assessment of health over 2 (that is, above cut 2) and calculate the log(odds) for assessing our own health as 4 or 5. This is the log(odds) for an assessment over 3, given that the assessment is over 2. A plot of this statistic against age group is shown in Figure 9.4. Again, we could consider fitting models of the form of (9.45), (9.46) and (9.47). We will refer to them as conditional odds models to distinguish them from the cumulative odds models. The parameters involving the 'cut' variable will now have a slightly different interpretation, but both cases can be regarded as 'nuisance' parameters since that is the effect of age group which is of real interest, Table 9.10(c) illustrate the rearrangement of the data in Table 9.10(a) to generate the required conditional odds. Agresti (1990, Chapter 9) and Aitkin *et al.* (1989) discuss these and other logistic models for ordered response variables. Lindsey (1989) provides details of GLIM macros for fitting them. We will not discuss the advantages of the different types of model, but will simply illustrate the use of the conditional odds models since they are readily fitted using current versions of GLIM.

Returning to Table 9.9, this data has been rearranged for consideration of conditional odds models in Table 9.11. Starting with the minimal model

$$\ln(\text{odds}) = \text{Cut} \tag{9.49}$$

Table 9.11 Conditional proportions obtained from the data in Table 9.9

Sex	Age group	Cut	No. above cut	Total
1	1	1	431	702
1	1	2	198	431
1	1	3	47	198
1	1	4	4	47
1	2	1	375	618
1	2	2	190	375
1	2	3	50	190
1	2	4	6	50
1	3	1	527	770
1	3	2	292	527
1	3	3	74	292
1	3	4	14	74
1	4	1	230	334
1	4	2	143	230
1	4	3	49	143
1	4	4	11	49
1	5	1	93	114
1	5	2	55	93
1	5	3	20	55
1	5	4	3	20
2	1	1	529	775
2	1	2	287	529
2	1	3	76	287
2	1	4	6	76
2	2	1	540	762
2	2	2	303	540
2	2	3	87	303
2	2	4	9	87
2	3	1	782	1005
2	3	2	502	782
2	3	3	169	502
2	3	4	24	169
2	4	1	335	427
2	4	2	234	335
2	4	3	87	234
2	4	4	16	87
2	5	1	202	245
2	5	2	149	202
2	5	3	66	149
2	5	4	24	66

we obtain a deviance of 281.90 with 36 degrees of freedom. This is equivalent to the log-linear model in (9.44). If we now fit

$$\ln(\text{odds}) = \text{Cut} + \text{Age} + \text{Sex} \tag{9.50}$$

a deviance of 28.91 with 31 degrees of freedom is obtained. The parameter estimates for this model are given in Table 9.12. Equation (9.50) is simply the description of eight parallel lines (log(odds) plotted against age group for each sex and cut) and provides a parsimonious description of the way age and sex jointly influence subjects perception of their own health.

Table 9.12   Parameter estimates for equation (9.42) fitted to the data in Table 9.11

Parameter	Estimate	Standard error
β_0	0·43	0·05
Sex (2)	0·33	0·04
Age (2)	0·07	0·06
Age (3)	0·38	0·05
Age (4)	0·55	0·06
Age (5)	0·83	0·08
Cut (2)	−0·56	0·04
Cut (3)	−1·77	0·05
Cut (4)	−2·67	0·11

As a second example in this section consider the data in Table 9.13. This is cross-classification of subjects with respect to two ordinal variables: parental socio-economic status and mental state. This table could be analysed through the use of the linear-logistic models described above (assuming mental state to be the response variable) but here we will discuss different classes of models, two variables to represent parental socio-economic status: a qualitative factor *PSOC* and a quantitative variate *PSCORE*. Similarly for mental state, consider a qualitative factor *MSTATE* and a qualitative variate *MSCORE* *PSCORE* and *MSCORE* are quantitative values assigned to the levels of *PSCOC* and *MSTATE*, respectively. These values can either be specified by the data analyst or they can be considered as model parameters to be estimated (see Agresti, 1990 Chapter 8).

The saturated log-linear model for the two-way table in Table 9.13 is the following:

$$PSOC + MSTATE + PSOC.MSTATE \tag{9.51}$$

If, in the two-way interactions we replace *PSOC* and *MSTATE* by *PSCORE* and *MSCORE*, respectively, we now have

$$PSOC + MSTATE + \beta_{pm}PSCORE.MSCORE \tag{9.52}$$

Now, if *PSCORE* and *MSCORE* are simply the values of *PSOC* and *MSTATE* respectively, then (9.52) is an example of a linear-by-linear association model. Its fit is indicated by a deviance of 9.89 with 14 degrees of freedom. The 15 estimated parameter values in (9.51) have been replaced by a single regression coefficient in (9.52).

In fitting both models (9.51) and (9.52) the estimates for the two-way interactions correspond to logarithms of odds ratios. For the saturated log-linear model:

$$PSOC(i).MSTATE(j) = \ln\left(\frac{m_{ij}m_{11}}{m_{1j}m_{i1}}\right) \tag{9.53}$$

For the linear-by-linear association model:

$$\beta_{pm} = \ln\left(\frac{m_{ij}m_{i+1,j+1}}{m_{i,j+1}m_{i+1,j}}\right) \tag{9.54}$$

Table 9.13 Association between state of mental health and parental socioeconomic status (from Srole *et al.*, 1962)

Mental state	Parental socio-economic status					
	A	B	C	D	E	F
Well	64	57	57	72	36	21
Mild symptom formation	94	94	105	141	97	71
Moderate symptom formation	58	54	65	77	54	54
Impaired	46	40	60	94	78	71

where the right-hand side of this expression is defined to be logorithms of the local odds ratio which, in this case, is constrained to be equal for all values of i and j. For the data in Table 9.13 β_{pm} is estimated as 0.09 (standard error 0.02).

Exercises

9.1 Explore the structure of the data in Table 9.3 using log-linear models. Compare your results with the analysis given in Section 9.3.

9.2 Analyse the counts in Table 7.2 using log-linear models (including the linear-by-linear association model). Then, taking 'suicide intent' as a response variable, re-analyse the data using an ordinal regression model. Compare your results.

9.3 After appropriate grouping of the counts, analyse the data given in Table 8.16 using log-linear models. These data were analysed using linear regression in Exercise 8.6, but the variable 'number of symptoms' can also be treated as if it were an ordinal response. Re-analyse these counts using a regression model for ordinal data. How does this analysis improve your understanding of counted data of this type?

9.4 Use linear-by-linear association models to explore the interactions between sex, age-group and health status in the data in Table 9.9. Compare your results with those arising for the use of ordinal regression models (Section 9.5).

9.5 Discuss the relative merits of the use of log-linear and linear-logistic models in the analysis of contingency tables.

9.6 Table 9.14 gives the results of a study of the relationship between car size and car accident injuries (Kihlberg *et al.* 1964). Accidents were classified according to type of accident, severity of accident, and whether or not the driver was ejected. Using 'severity' as the response variable, fit and interpret a linear-logistic model to these counts.

9.7 Analyse and interpret the counts given in Table 9.8.

Table 9.14 The effect of car weight, accident type and ejection of driver on the severity of accidents (Kihlberg *et al.,* 1964)

Car weight	Driver ejected	Accident type	No. severe	No. not severe
Small	No	Collision	150	350
Small	No	Rollover	112	60
Small	Yes	Collision	23	26
Small	Yes	Rollover	80	19
Standard	No	Collision	1022	1878
Standard	No	Rollover	404	148
Standard	Yes	Collision	161	111
Standard	Yes	Rollover	265	22

10

Models for Rates and Survival Times

10.1 Introduction

This chapter illustrates the use of log-linear models for the analysis of rates and for the analysis of survival times. Other models for the analysis of survival times are also briefly discussed. In the case of rates, the measurements to be analysed are the number of events of some specified type (deaths, accidents, marriages, divorces, memory recalls, faults, and so on) divided by a relevant baseline measure (total time of observation, length of cable or yarn inspected, geographical area of study, and so on). Survival times are the observed times from the intitiation of a process of interest (birth or time of marriage, for example) and the occurrence of an event of interest (death or divorce, for example). Table 10.1, for example, shows remission times, in weeks, of two ways groups of leukaemia patients, one of which had been treated with an active drug (6-mercaptopurine) and the other with a placebo. Note that these remission times, in common with many examples of this type of data, are censored. This indicates that the remission time is known to be longer than that recorded, but the exact time of recurrence or whether there was subsequent recurrence of the illness is not known. Possible reasons for censoring in this example could be death from a cause other than leukaemia or that the patients were still in remission at the end of the trial.

Table 10.2 presents memory recall times for an experiment involving a single male subject. These data arise from a study in which the subject was told that he would be presented with a series of stimulus words on a computer screen, one at a time, and each word would be preceded by an instruction on the screen asking him to recall either a pleasant or an unpleasant memory associated with the next word. Requests to recall pleasant and unpleasant memories alternated in the sequence, and successful recall of a memory was acknowledged by pressing the bar on the computer keyboard. If the subject was unable to recall a memory within fifteen seconds, the experimenter moved on to another stimulus word. In other words, the memory recall times were censored at fifteen seconds, and where appropriate these are represented by '15+' in Table 10.2. The latencies for twenty-four pleasant memories are to be compared with those for the same number of unpleasant ones.

Table 10.1   Times of remission (in weeks) of leukaemia patients (taken with permission from Freirich *et al.*, 1963)

Sample 1	6*	6	6	6	7	9*	10*
(drug)	10	11*	13	16	17*	19*	20*
	22	23	25*	32*	32*	34*	35*
Sample 2	1	1	2	2	3	4	4
(placebo)	5	5	8	8	8	8	11
	11	12	12	15	17	22	23

Total observation time for Sample 1 = 359 wks.
Number of recurrencies of illness in Sample 1 = 9.
Rate of remission in Sample 1 = 0·03 per wk.
Total observation time for Sample 2 = 182 wks.
Number of recurrencies of illness in Sample 2 = 21.
Rate of remission in Sample 2 = 0·12 per wk.

* Censored; that is, the remission time is known to be longer than that recorded, but the exact time of remission is not known.

Table 10.2   Memory recall times for one male subject (from Dunn and Master, 1982)

Pleasant	Unpleasant	$\hat{S}(t)$
1·07	1·45	0·96
1·17	1·67	0·92
1·22	1·90	0·88
1·42	2·02	0·83
1·63	2·32	0·79
1·98	2·35	0·75
2·12	2·43	0·71
2·32	2·47	0·67
2·56	2·57	0·63
2·70	3·33	0·58
2·93	3·87	0·54
2·97	4·33	0·50
3·03	5·35	0·46
3·15	5·72	0·42
3·22	6·48	0·38
3·42	6·90	0·33
4·63	8·68	0·29
4·70	9·47	0·25
5·55	10·00	0·21
6·17	10·93	0·17
7·78	15+	0·13/0·17*
11·55	15+	0·08/0·17*
15	15+	0·08/0·17*
15+	15+	0·08/0·17*

* The first number refers to pleasant memories, the second to unpleasant ones.

Not all 'survival' times correspond to a time prior to the occurrence of a unique event. A patient can have more than one bout of illness with periods of remission between them; and a subject can have more than one period of unemployment, for example, or get married several times in a lifetime. One can also study durations to events of qualitatively different types within the same study. A patient, for example, may die from the result of several competing causes such as a traffic accident, a heart attack, cancer, and so on. A period of employment might end for several reasons. An employee for example might be made redundant, be dismissed for incompetence or retire because of age or disability. He may also simply decide to change his job.

A comprehensive description of the methods available for modelling survival times and other event history data is well beyond the scope of this chapter. Here we will illustrate some general ideas using relatively simple examples. For further details readers are referred to Lee (1980), Kalbfleisch and Prentice (1980), Allison (1984), and Cox and Oakes (1984).

10.2 Log-Linear Models for Rates

Table 10.1 indicates that the rate of relapse for the drug-treated leukaemia patients is about 0.03 per week. The corresponding rate for the placebo-control is 0.12 per week. Consider a log-linear model of the following form:

$$\ln\left(\frac{n_i}{t_i}\right) = \beta_0 + \text{Drug}(i) \tag{10.1}$$

where n_i is the expected number of relapses of drug group i (where $i = 1$ for placebo and $i = 2$ for the active drug) and t_i is the corresponding total observation time. This equation is equivalent to

$$\ln(n_i) = \beta_0 + \text{Drug}(i) + \ln(t_i) \tag{10.2}$$

We can now fit a log-linear model to the N conditional on the observation times t_i. The term $\ln(t_i)$, which is regarded as a constant and not a parameter to be estimated, is called an *offset*. Constraining Drug(1) to be zero as in earlier models, Drug(2) is then a measure of the logarithm of the ratio of observed rates for the two patient groups. To test whether this term is significantly different for zero (equal rates), we simply delete it from the model. Similarly, for the memory recall data in Table 10.2 we could fit a similar pair of models, replacing Drug(i) in (10.2) with PLEASURE(i) where $i = 1$ for unpleasant memories and $i = 2$ for pleasant ones.

The results of the analysis of these two data sets are shown in Tables 10.3(a) and 10.3(b), respectively.

A third example is given in Table 10.4. Table 10.4(a) contains the results for the study of survival after two different types of valve replacement as a function of the patient's age group (Cohn *et al.*, 1981). These data have also been analysed by Laird and Olivier (1981) and by Agresti (1990). Table 10.4(b) shows the results of fitting alternative models:

$$\ln(n_{ij}) = \beta_0 + \text{Age}(i) + \text{Type}(j) + \ln(t_{ij}) \tag{10.3}$$

$$\ln(n_{ij}) = \beta_0 + \text{Age}(i) + \ln(t_{ij}) \tag{10.4}$$

$$\ln(n_{ij}) = \beta_0 + \text{Type}(j) + \ln(t_{ij}) \tag{10.5}$$

and

$$\ln(n_{ij}) = \beta_0 + \ln(t_{ij}) \tag{10.6}$$

Table 10.3   Log-linear models

(a) Rates of leukaemia relapse
$$\text{Model}: \ln(n_i) = \beta_0 + \text{Drug}(i) + \ln(t_i)$$
$$G^2 = 0 \text{ with } 0 \text{ d.f.}$$

Parameter	Estimate	Standard error
β_0	$-2 \cdot 16$	$0 \cdot 22$
Drug (2)	$-1 \cdot 53$	$0 \cdot 40$

$$\text{Model}: \ln(n_i) = \beta_0 + \ln(t_i)$$
$$G^2 = 16 \cdot 49 \text{ with } 1 \text{ d.f.}$$

(b) Rates of memory recall
$$\text{Model}: \ln(n_i) = \beta_0 + \text{Pleasure}(i) + \ln(t_i)$$
$$G^2 = 0 \text{ with } 0 \text{ d.f.}$$

Parameter	Estimate	Standard error
β_0	$-2 \cdot 04$	$0 \cdot 22$
Pleasure (2)	$0 \cdot 46$	$0 \cdot 31$

$$\text{Model}: \ln(n_i) = \beta_0 + \ln(t_i)$$
$$G^2 = 2 \cdot 20 \text{ with } 1 \text{ d.f.}$$

---

where n_{ij} is the expected number of deaths for age group i ($i = 1$ if less than 55, $i = 2$ if 55 or over) and heart valve j ($j = 1$ for aortic $j = 2$ for mitral). Age(i) is the effect of the ith age group and Type(i) the effect of the jth value type. The offset for these models is $\ln(t_{ij})$ where t_{ij} is the total observed exposure time for the ijth cell of the table. It will be left as an exercise for the reader to interpret these results.

The fitting of the above log-linear models is based on the assumptions of independence of the survival times and, if some subjects provide data for multiple events (times between successive accidents, for example), then modelling the number of events as Poisson variates will be invalid. The number of events will be more variable than expected for Poisson variates — an example of *over-dispersion*. Methods of dealing with over- or under-dispersion are beyond the scope of this text, but are dealt with in McCullagh and Nelder(1989). Other implicit assumptions made in fitting these log-linear models to event data are discussed in the next section.

10.3   Log-Linear Hazard Models

Returning to the recorded response latencies in Table 10.2 the first thing that might be of interest about such data is the survival function, which is defined as the probability that an individual latency, T, is longer than a given time t. That is,

$$S(t) = Pr(T > t) \tag{10.7}$$

The survival function $S(t)$ can be estimated by the proportion of latencies found to be longer than each of the observed recall times, t. Graphs of $S(t)$ against t for both pleasant and unpleasant memories are plotted in Figure 10.1. Such plots are known as survival curves. (Note that they are *step functions*, although they are often represented by smooth curves). In cases where there

Table 10.4 Death rates after heart valve replacement operations (after Cohn *et al.*, 1981)

(a) Data

Age group		Type of heart valve Aortic	Mitral
Young (< 55)	Deaths	4	1
	Total exposure time (months)	1259	2082
Old (55+)	Deaths	7	9
	Total exposure time (months)	1417	1647

(b) Log-linear models

Model: $\ln(n_{ij}) = \beta_0 + \text{Age } (i) + \text{Type } (j) + \ln(t_{ij})$
$G^2 = 3.22$ with 1 d.f.

Parameter	Estimate	Standard error
β_0	−6·31	0·51
Age (2)	1·22	0·51
Type (2)	0·33	0·44

Model: $\ln(n_{ij}) = \beta_0 + \text{Age } (i) + \ln(t_{ij})$
$G^2 = 3.79$ with 2 d.f.

Model: $\ln(n_{ij}) = \beta_0 + \text{Type } (j) + \ln(t_{ij})$
$G^2 = 9.89$ with 2 d.f.

Model: $\ln(n_{ij}) = \beta_0 + \ln(t_{ij})$
$G^2 = 10.04$ with 3 d.f.

Table 10.5 Product-limit estimation for the survival function for drug-treated leukaemia patients

Survival time (t_i)	No. at risk (n_i)	No. relapses (r_i)	$\left(1 - \dfrac{r_i}{n_i}\right)$*	$S(t_i)$**
6	21	3	0·8571	0·8571
7	17	1	0·9412	0·8067
10	15	1	0·9333	0·7529
13	12	1	0·9167	0·6902
16	11	1	0·9091	0·6275
22	7	1	0·8571	0·5378
23	6	1	0·8333	0·4481

* Conditional probability of surviving at t_i given that survived to t_{i-1}
** $\Pi_j(1 - r_j/n_j)$ = product of conditional probabilities

is censoring of the survival types, an in the leukaemia data in Table 10.1, the survival function will need to be estimated to take this into account. The most usual method is that described by Kaplan and Meier (1958) — the *product limit estimator*. This method is illustrated for the drug-treated leukaemia patients in Table 10.5. Further details can be found in Armitage and Berry (1987).

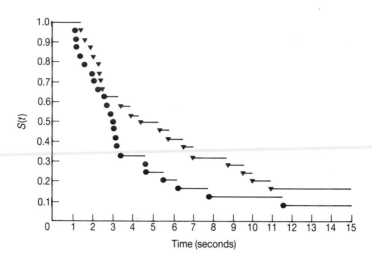

Figure 10.1

Table 10.6 gives some other examples of the way in which the data in Table 10.2 can be summarised. For each type of memory are given median recall times. Clearly, since the memory recall times are censored, the arithmetic mean of the recall times is not a summary statistic that one can sensibly compute. As a measure of variation of the latencies the interquartile range or midspread is given for the two types of response. Table 10.6 also gives the number of observed latencies that are greater than 2, 5, 10 and 15 seconds, respectively. The choice of these times is to some extent arbitrary, but is obviously influenced by the shape of the survival curves such as those in Figure 10.1. As would be expected, these counts convey very similar information to the curves in Figure 10.1. Both Table 10.6 and Figure 10.1 show clearly that the latencies for unpleasant memories are, on average, longer than those for the pleasant ones. The grouping of latencies in Table 10.6 has been done to aid the visual interpretation of the data and to illustrate how models for grouped survival times can be approached. The latter is the subject of Section 10.5, but it is worthwhile noting here the similarity of the grouped data in Table 10.6 with that for the ordinal or counted response variables in Chapter 9.

Here we introduce models that are applicable to latencies measured on a continuous time scale. Having introduced the survival function, we will now introduce the corresponding hazard rate, $h(t)$. This is defined as follows:

$$h(t) = \lim_{\delta t \to 0} \frac{P(T \leq t + \delta t | T > t)}{\delta t} \qquad (10.8)$$

In words, the hazard rate is the instantaneous recall rate at time t. The simplest form that the hazard function can have is to be constant, that is $h(t) = \lambda$, for all values of t. This is the characteristic of the negative exponential distribution with probability density function, $f(t)$, given by

$$f(t) = \lambda \exp(-\lambda t), \quad t > 0 \qquad (10.9)$$

Table 10.6   Summary statistics for the latencies in Table 10.2

Time (t)	Number of latencies greater than or equal to t	
	Pleasant	Unpleasant
2	18	21
5	6	12
10	3	6
15	2	4

Median response time for pleasant memories: 2·97 seconds
Interquartile range: 1·98–4·70 seconds
Median response time for unpleasant memories: 4·33 seconds
Interquartile range: 2·35–9·47 seconds

and corresponding cumulative distribution function, $F(t)$, given by

$$F(t) = 1 - \exp(-\lambda t), \quad t > 0 \tag{10.10}$$

Here the survival function, $S(t)$, has the form $S(t) = \exp(-\lambda t)$, so that a plot of $\ln(S(t))$ against t will be a straight line with slope $-\lambda$.

Consider the situation where we may wish to model some function of the latencies for the two types of memory. It is customary to model the hazard rate rather than the expected value of survival or recall time. One obvious choice is a log-linear model for the hazard rates. For the example in Table 10.2, we might have

$$\ln h(t) = \beta_0 + \text{PLEASURE}(i) \tag{10.11}$$

where β is a constant and PLEASURE(i) is the effect of unpleasant ($i = 1$) or pleasant ($i = 2$) memories. Note that this model implies a constant ratio of hazards for the two memory types. It is a specific example of a more general family of proportional hazards models (see below).

The results of fitting model (10.11) to the memory data are given in Table 10.7. Note that the parameter estimates, their standard errors, and the change in deviance due to memory type, are all identical to those given for the log-linear rate model in Table 10.3(b). The two approaches are equivalent and the assumptions implicit in fitting the models in Section 10.2 are those explicitly made above. One advantage of fitting the response times individually, however, is that the results enable one to judge the adequacy of the model. The deviance statistics given in Table 10.7 for example, indicate that the constant hazard assumption is reasonable for the memory data in Table 10.2. For further details, the reader is referred to Agresti (1990) and to Aitkin and Clayton (1980).

For the memory recall data presented in Table 10.2 a preliminary explanatory analysis (a plot of $\ln(S(t))$ against t) indicates that a log-linear exponential model describes the data adequately, except that there might be a period of up to about one second during which the hazard function appears to be zero. That is, there is a short period during which it is impossible to recall a memory (the guarantee time). This period will be symbolised by G. The

Table 10.7 Results of fitting negative exponential survival models to the memory data
(Table 10.2)

Model: $\ln(h(t)) = \beta_0 + \text{Pleasure}(i)$
$G^2 = 2 \cdot 20$ with 1 d.f.

Parameter	Estimate	Standard error
β_0	$-2 \cdot 04$	$0 \cdot 22$
Pleasure (2)	$0 \cdot 46$	$0 \cdot 31$

Model: $\ln(h(t = \beta_0$
$G^2 = 52 \cdot 81$ with 47 d.f.
Change in $G^2 = 2 \cdot 20$ with 1 d.f.

modified exponential distribution has a hazard given by

$$h(t) = 0 \quad \text{for} \quad 0 \le t < G$$

$$h(t) = \lambda \quad \text{for} \quad t > G \tag{10.12}$$

The equivalent survival function is

$$S(t) = \exp(-\lambda(t - G)) \qquad \text{for } t > G$$

$$S(t) = 1 \quad \text{for} \quad t < G \tag{10.13}$$

There are many other ways in which the hazard function might change with
time. In the context of machine failure, for example, the hazard is likely to
increase with time due to the process of wearing of essential components. In
other contexts the hazard might decrease with time. In the memory recall
experiment, for example, the subject might slowly give up trying to recall a
memory. A generalisation of the exponential distribution to allow for changes
in the hazard rate with time is the Weibull distribution. In this case the hazard
function is

$$h(t) = \lambda\alpha(\lambda t)^{\alpha-1} \quad \text{for } t > 0 \text{ and } \alpha > 0 \tag{10.14}$$

This reduces to the exponential when $\alpha = 1$. Another alternative is the
Gompertz distribution characterised by

$$h(t) = \lambda\alpha^t \quad \text{for} \quad t > 0 \tag{10.15}$$

which again reduces to the exponential distribution when $\alpha = 1$.

In the analysis of life-tables we find that a plot of hazard against age is
U-shaped. Here the hazard (mortality rate) is relatively high for young babies,
decreases for older children and young adults, and then rises steeply in old age.
As another example, if one considers patterns of divorce, it is easy to imagine
a situation where the hazard might increase for the first few years of married
life, and then slowly decrease. Each phenomenon has its own characteristic
hazard distribution, and many of these distributions cannot be described using
the above models. Such problems are overcome in Cox's proportional hazards
model (Cox, 1972). In this model, for example, equation (10.11) is replaced by

$$\ln h(t) = \alpha(t) + \beta_0 + \text{PLEASURE}(i) \tag{10.16}$$

where $\alpha(t)$ is an arbitrary function of time. This is a proportional hazards
model because my two memories at any given point in time still have a

constant ratio of hazards. Because the baseline hazard function, $h(t)$, does not have to be specified explicitly in the Cox model, this proportional hazards model is often referred to as being non-parametric or, more correctly, as semi- or partially parametric. For details the reader is referred to Kalbfleisch and Prentice (1980) and to Cox and Oakes (1984). The use of the Cox model will not be illustrated here, but the equivalent model for grouped survival times will be used in the following section.

General strategies for modelling survival times are in many ways similar to those described in Chapters 8 and 9. Stepwise selection procedures, for example, can be used in a similar way. Kay (1984) reviews the analysis of residuals and other goodness-of-fit statistics for these models. Of particular interest, however, are preliminary plots of the data to check, for example, whether the proportional hazards assumption holds or whether the hazard rate is a simple function of time. We have already mentioned a plot of $\ln(S(t))$ against t for the exponential distribution. Consider the integrated or cumulative hazard function, $H(t)$, where it can be shown that

$$H(t) = -\ln S(t) \tag{10.17}$$

Details of its derivation can be found, for example, in Cox and Oakes (1984). Diagnostic plots using estimated values of $H(t)$ can be useful in checking model assumptions. As has already been stated a plot of $H(t)$ against t gives a straight line for the exponential distribution. A Gompertz distribution implies a straight line in a plot of $\ln(H(t))$ against t and, similarly, a Weibull distribution implies a straight line in a plot of $\ln(H(t))$ against $\ln(t)$. Whatever the shape of the $H(t)$ versus t plot, the proportional hazards assumption for the two types of memory in Table 10.2, for example, implies that the plots for the two groups will be parallel.

The proportional hazards assumption can also be checked as part of the modelling procedure itself. Model (10.11), for example, might be modified to include an interaction term involving memory type and observation time. A significance test for this interaction is then essentially a test of the proportional hazards assumption. A similar change could also be introduced to the Cox model in equation (10.16).

10.4   Models for Grouped Survival Times

In many situations the exact time of occurrence of a crucial event is not known. A respondent may recall, for example, the year in which he was made redundant by his employer but not the date. In follow-up studies, investigators might simply ask whether there has been an episode of illness since the previous interview and again not obtain information on the precise timing. Finally, even though exact times may have been collected in a study of event histories the investigators may decide to simplify the data by first aggregating the data into convenient time intervals. This might be done, particularly for large samples, for computational convenience and ease of interpretation of a table of data. An example of data of this type is given in Table 10.8.

Table 10.8 arose from the analysis of records of psychiatric illness in a single general practice over the 20-year period 1957 to 1976 (Dunn and Skuse, 1981). Data from the first and second decades of the study were analysed separately. Table 10.8 illustrates one aspect of the data from the first decade. This table

Table 10.8   The effects of age and sex on the first episodes of depression (after Dunn and Skuse, 1981)

Year	Sex	Age group	No. diagnosed as depressed for first time in study	No. at risk
1962	Male	1	0	56
1963	Male	1	0	56
1964	Male	1	0	56
1965	Male	1	1	56
1966	Male	1	0	55
1962	Male	2	3	193
1963	Male	2	4	190
1964	Male	2	4	186
1965	Male	2	7	182
1966	Male	2	3	175
1962	Male	3	5	213
1963	Male	3	3	208
1964	Male	3	3	205
1965	Male	3	5	202
1966	Male	3	1	197
1962	Male	4	1	67
1963	Male	4	1	66
1964	Male	4	1	65
1965	Male	4	1	64
1966	Male	4	0	63
1962	Female	1	0	28
1963	Female	1	1	28
1964	Female	1	0	27
1965	Female	1	0	27
1966	Female	1	1	27
1962	Female	2	10	142
1963	Female	2	8	132
1964	Female	2	8	124
1965	Female	2	10	116
1966	Female	2	7	106
1962	Female	3	13	126
1963	Female	3	11	113
1964	Female	3	2	102
1965	Female	3	8	100
1966	Female	3	5	92
1962	Female	4	9	84
1963	Female	4	7	75
1964	Female	4	2	68
1965	Female	4	0	66
1966	Female	4	5	66

also illustrates another aspect of censoring of survival data. The records began in 1957 and are left censored (as opposed to right censored as described, for example, in Tables 10.1 and 10.2). There is no information on episodes of depression prior to the beginning of the study and a patient free of depression at the beginning of the period of the data collection may have been free since birth or may have had a bout of depression two years earlier. To avoid this problem the data in Table 10.8 contain only information on patients known to be free of depression in the first five years of the study. A 'first episode' of depression was then arbitrarily defined as the first episode recorded in the period 1962 to 1966 given that the patients had no recorded episodes in the period 1957 to 1961. Patients who were ill in the period prior to 1962 are not included in Table 10.8. The patients are categorised according to their sex and age in 1966, four age groups being used:

$$(1)\ 20 - 30;\quad (2)\ 31 - 45;\quad (3)\ 46 - 60;\quad (4)\ \text{over } 60$$

If, as might have been more sensible, the patients had been put into the four age groups at each year in the period 1962 to 1966, then some of the patients would have obviously changed from one age group to the next during the course of the follow-up period. This would then have provided an example of a time-dependent explanatory factor, as if the subjects' exact ages had been used as a continuous covariate, a time-dependent or time-varying covariate. This aspect of the analysis of a longitudinal data file will be discussed in the next chapter.

Here we define the discrete-time analogue of the hazard rate discussed in the previous section. Here we are simply interested in the conditional probability of having an episode of depression in a given year given that the patient has been free of depression in the period up to that year. For the year 1962, for example, this is estimated by

Number depressed in 1962 but not in 1957–61/ Number free of depression in 1957–61

If right censoring were present through deaths or emigration of patients then the above proportion could be modified by changing the bottom line to "number at risk in 1962." In Table 10.8, however, this problem has been avoided because it deals only with the subset of patients who were on the general practice register for the full twenty years of the study. Similarly, for 1963, the conditional probability is estimated by

Number depressed in 1963 but not in 1957–62/ Number free of depression 1957–62

The obvious choice for a linear model for these conditional probabilities has the same form as the conditional odds models for ordinal responses in the last chapter. The arbitrary change in hazard throughout the five years of follow-up is modelled through the effect of a qualitative factor Year(k), $k = 1$ to 5, which is analogous to the cut parameter of Section 9.6. Assuming that there are no interactive effects, one could start by fitting

$$\ln(\text{odds}) = \beta_0 + \text{Sex}(i) + \text{Age}(j) + \text{Year}(k) \qquad (10.18)$$

where the labelling of the parameters should be clear for the context. To constrain the hazards to be constant over time, this model could then be replaced by

$$\ln(\text{odds}) = \beta_0 + \text{Sex}(i) + \text{Age}(j) \qquad (10.19)$$

and the effect of Year(k) assessed by the increase in deviance produced.

Linear-logistic models such as those in (10.18) and (10.19) are obviously models incorporating the assumption of proportional odds rather than proportional hazards. If the proportional hazards model were to hold for survival

Table 10.9 Survival models for first episodes of depression

Model	d.f.	Deviance for logistic link	Deviance for compl. log-log link
(a) Sex+Age+Year	31	32·35	32·36
(b) Sex+Age	35	39·37	39·39
(c) Sex	38	53·56	53·56

Parameter estimates for model (a)

Parameter	Logistic link estimate	s.e.	Complementary log-log link estimate	s.e.
β_0	−5·39	0·61	−5·39	0·60
Sex (2)	1·38	0·19	1·36	0·18
Age (2)	1·69	0·59	1·66	0·59
Age (3)	1·62	0·60	1·60	0·59
Age (4)	1·48	0·61	1·46	0·61
Year (2)	−0·09	0·24	−0·09	0·23
Year (3)	−0·61	0·28	−0·60	0·27
Year (4)	−0·08	0·24	−0·08	0·24
Year (5)	−0·42	0·27	−0·41	0·26

Parameter estimates for model (b)

Parameter	Logistic link estimate	s.e.	Complementary log-log link estimate	s.e.
β_0	−5·61	0·59	−5·60	0·59
Sex (2)	1·39	0·18	1·37	0·18
Age (2)	1·70	0·59	1·67	0·59
Age (3)	1·63	0·60	1·61	0·59
Age (4)	1·48	0·61	1·47	0·61

data prior to grouping then it can be shown that the appropriate link function for the grouped data should have the following form (Prentice and Gloeckler, 1978):

$$\ln(-\ln(1 - p)) \qquad (10.20)$$

where p is the expected value of the appropriate conditional probability. In GLIM this link function is referred to as the complementary log-log link. The grouped-time equivalent to the Cox proportional hazards model for the data in Table 10.8 is therefore

$$\ln(-\ln(1 - p)) = \beta_0 + \text{Sex}(i) + \text{Age}(i) + \text{Year}(k) \qquad (10.21)$$

In most situations the choice between logistic and complementary log-log links will have little practical significance, and will be more dependent on the investigator's theoretical position. Here we present the results of using both links — see Table 10.9.

Exercises

10.1 From the definitions of $h(t)$ and $S(t)$ in Section 10.3 show that

$$h(t) = \frac{f(t)}{F(t)}$$

where $f(t)$ is the probability distribution function and $F(t)$ the cumulative distribution function (where, by definition, $S(t) = 1 - F(t)$), and that

$$f(t) = h(t)\exp(-H(t))$$

where the cumulative hazard function, $H(t)$, is defined by

$$H(t) = \int_0^t h(u)du$$

10.2 Demonstrate that for survival times distributed according to the exponential distribution a plot of $H(t)$ against t will yield a straight line. Show that for the Gompertz distribution a straight line will be obtained in a plot of $\ln(H(t))$ against t. Similarly, show that for the Weibull distribution a straight line will be obtained in a plot of $\ln(H(t))$ against $\ln t$.

10.3 Fit survival models to the data in Tables 10.1 and 10.2 assuming that the survival times in each group follow a Weibull distribution.

10.4 Fit survival models to the data in Table 10.2 using the Cox proportional hazards model. Returning to the use of exponential models, assess whether there is a need to introduce a guarantee or minimum response time in models for these data (see Lee, 1980).

10.5 Analyse the data in Table 10.10. This shows survival time (i.e. time to death) in two groups of leukaemia patients in weeks and their white blood cell count (WBC).

10.6 Table 10.11 shows the results of fitting linear-logistic survival models to the depression data in Table 10.8 and to the corresponding node for the following data (1967–76). Both models include a Sex.Age interaction. Interpret these results, paying particular attention to changes that may have occurred over the 10-year period between one decade and the next (see Dunn and Skuse 1981).

Table 10.10 Survival time and white blood cell count (WBC)
(Reproduced with permission from Feigl and Zelen (1965))

AG-positive (N = 17) WBC	Survival time (wks)	AG-negative (N = 16) WBC	Survival time (wks)
2 300	65	4 400	56
750	156	3 000	65
4 300	100	4 000	17
2 600	134	1 500	7
6 000	16	9 000	16
10 500	108	5 300	22
10 000	121	10 000	3
17 000	4	19 000	4
5 400	39	27 000	2
7 000	143	28 000	3
9 400	56	31 000	8
32 000	26	26 000	4
35 000	22	21 000	3
100 000	1	79 000	30
100 000	1	100 000	4
52 000	5	100 000	43
100 000	65		

Table 10.11 Parameter estimates for linear-logistic survival models for 'first' episodes of depression (Dunn and Skuse, 1981)

Parameter	1962–66 Est.	s.e.	1972–76 Est.	s.e.
β_0	−5·41	1·00	−0·36	0·29
Year (2)	−0·09	0·24	0·02	0·20
Year (3)	−0·61	0·28	−0·20	0·22
Year (4)	−0·08	0·24	0·10	0·21
Year (5)	−0·02	0·27	−0·62	0·26
Age (2)	1·86	1·02	1·08	0·35
Age (3)	1·54	1·03	0·44	0·31
Age (4)	1·24	1·12	0·21	0·33
Sex (2)	1·41	1·23	0·75	0·35
Age (2).Sex (2)	−0·25	1·25	−1·06	0·54
Age (3).Sex (2)	0·12	1·26	−0·60	0·44
Age (4).Sex (2)	0·28	1·34	0·39	0·43

11

Analysis of Repeated Measures

11.1 Introduction

In this chapter we will concentrate on the use of modelling techniques for the investigation of the results of experiments or surveys in which the behaviour of an individual, or a group of individuals, is monitored over time. Such data arise in many areas of biology, medicine, psychology and sociology, for example. Almost everyone will be familiar with routinely collected measurements of children's weight, height or intellectual development. Equally familiar will be the widespread use of economic indicators such as exchange rates, rates of inflation, average share prices, and so on. Another example is use of political opinion polls to chart changes in voting intentions.

Table 11.1 shows a set of data arising in the investigation of the approach avoidance nature of an autistic child's social behaviour. The child's behavioural state was recorded after consecutive intervals of about two seconds during a period of time in which he was engaged in a continuous one-to-one activity with his teacher. The child's behavioural repertoire was grouped into three exhaustive and mutually exclusive states which were:

(a)   activities indicating a positive approach to his teacher (coded as 1),

(b)   activities indicating resistance or avoidance of the teacher (coded as 3),

(c)   inactivity or solitary preoccupation that gave no indication that the child might be aware of the other person (coded as 2).

(Note that the three coded behavioural states can be regarded as ordinal measures of the child's interpersonal responsiveness.)

More typical of the type of data to be discussed in this chapter are those illustrated by the use of a panel survey. A sample of individuals (the panel) is interviewed at regular intervals (in January of each of five successive years, for example) and data concerning their current state or state over the previous interval is recorded. Well-known examples include questions on voting intentions in British and American elections. Table 11.2 illustrates the use of longitudinal data arising from the annual summary of medical records. Here the subject is coded as having suffered from one or more bouts of depression (D) in the

Table 11.1   An autistic child's behavioural sequence
(from Dunn and Clark, 1982)

```
1113231112322332232322222222111212112122
2233222222122222223322222322221122111211
1232222222221222222231121111111221321111
1111212111122212222113111122222222222111
1212222222222222233311111111111222222212
2211111111111112222111133222222222222222
```

Table 11.2   Hypothetical longitudinal data file recording the presence (D) or absence (W) of bouts of depression in each of five successive years of a medical survey

Subject	Age in Yr 1	Sex		Presence of depression			
		Yr 1	Yr 2	Yr 3	Yr 4	Yr 5	
1	23	Male	W	W	W	W	D
2	62	Female	D	D	W	D	D
3	43	Female	W	W	W	W	W
4	29	Female	W	W	D	D	D
5	42	Male	D	D	W	W	W
⋮	⋮	⋮	⋮	⋮	⋮	⋮	⋮

previous year or as being well (W) throughout the year (see for example Dunn and Skuse, 1981).

Table 11.3 illustrates a data set arising from a repeated measures experiment. Here patients, with either mild or severe symptoms, were treated with either of two drugs (the standard treatement is the new one) and their progress was checked at three time points. Here the interest is in the comparisons of the profiles of three responses for each of the four combinations of diagnosis and treatment. More typically, a repeated measures experiment will involve the use of quantitative rather than categorical outcome measures. Kenward (1987), for example, gives an example where outcome is measured by the repeated measurement of the weight of 60 calves (eleven measurements for each calf) in an experimental trial on the control of intestinal parasites.

One type of repeated measures experiment that will not be considered in the present chapter is one in which each subject receives more than one of the alternative treatments or experimental conditions (see, for example, Hand and Taylor, 1987 or Jones and Kenward, 1989). There will be situations, however, where possible explanatory factors, or covariates, might change during the course of an experiment or survey. One example is time itself and another is the age of the respondent (see Table 10.3, for example). On the whole we will be concerned with the analysis of what Laird and Ware (1982) have classified as growth curves: situations in which natural development or the ageing process of the individual is monitored without intervention, but where comparisons are often made between different groups. We will, however,

Table 11.3 Cross-classification of responses at three times (N = normal, A = abnormal) by diagnosis and treatment
(Reprinted with permission from the Biometric Society (Koch *et al.*, 1977)

Diagnosis	Treatment	Responses at three times*							
		NNN	NNA	NAN	NAA	ANN	ANA	AAN	AAA
Mild	Standard	16	13	9	3	14	4	15	6
Mild	New Drug	31	0	6	0	22	2	9	0
Severe	Standard	2	2	8	9	9	15	27	28
Severe	New Drug	7	2	5	2	31	5	32	6

* 1 week, 2 weeks, and 4 weeks.

Table 11.4 Transition counts for Markov chains

(a) First order			Diagnosis in ith year		Total
			D	W	
	Diagnosis in	D	n_{DD}	n_{DW}	n_{D+}
	(I−1)th year	W	n_{WD}	n_{WW}	n_{W+}

(b) Second order		Diagnosis in ith year		Total
Diagnosis in (i−2)th year	Diagnosis in (i−1)th year	D	W	
D	D	n_{DDD}	n_{DDW}	n_{DD+}
	W	n_{DWD}	n_{DWW}	n_{DW+}
W	D	n_{WDD}	n_{WDW}	n_{WD+}
	W	n_{WWD}	n_{WWW}	n_{WW+}

consider situations where the different groups of subjects might be experimental conditions imposed by the investigator.

11.2 Markov Chains and Autoregressive Models

Initially, let us consider a hypothetical investigation in which a patient is categorised as depressed (D) or well (W) at each of a number of equally spaced points in time, say yearly. This is analogous to the situation displayed in Table 11.2. If there are many patients being observed in this longitudinal study then the pattern of diagnosis for two consecutive years can be summarised as in Table 11.4(a). This is a table of transition counts, the transition of interest being that from the diagnosis in the $(i-1)$th year to that in the ith year, the previous states being ignored. Table 11.4(b) shows a similar table for the pattern of diagnosis for three consecutive years.

Concentrating for the moment on Table 11.4(a), we may use the entries in this table to estimate the probability of being depressed at the ith diagnosis, given a diagnosis of depression at the immediately preceding diagnosis. This estimate is given simply by $\frac{n_{DD}}{n_{D+}}$. Similarly, the probabilities of being diagnosed as depressed at the ith examination, given that the subject is well at the

Table 11.5 Transition probabilities for Markov chains

(a) First order			*Diagnosis in ith year*		*Total*
			D	W	
	Diagnosis in	D	P_{DD}	P_{DW}	1
	(i–1)th year	W	P_{WD}	P_{WW}	1

(b) Second order		*Diagnosis in ith year*		*Total*
Diagnosis in	*Diagnosis in*	D		
(i–2)th year	*(i–1)th year*		W	
D	D	P_{DDD}	P_{DDW}	1
	W	P_{DWD}	P_{DWW}	1
W	D	P_{WDD}	P_{WDW}	1
	W	P_{WWD}	P_{WWW}	1

previous examination, is estimated by $\frac{n_{WD}}{n_{W+}}$. Four such transition probabilities can be estimated from Table 11.4(a) and these are shown in Table 11.5(a). Note that the sum of P_{DD} and P_{DW} is unity, as is that of P_{WD} and P_{WW}. There are, therefore, only two independent transition probabilities to be estimated. Here we shall consider P_{WD}, the probability of becoming depressed between consecutive diagnosis, and P_{DD}, the probability of remaining depressed. (The transition probabilities corresponding to the frequencies in Table 11.4(b) arc shown in Table 11.5(b). Note that here there are four independent transition probabilities that can be estimated.)

The estimates of the transition probabilities given above are based on the assumption that these do not change over time; that is, that the process is stationary. For, such a sequence, if the outcome of a particular diagnosis is dependent only on the immediately prior diagnosis, the process of change in patterns of diagnosis is called a first-order *autoregressive process* or a first-order *Markov chain* (or simply Markov chain). If a diagnosis is dependent on the two preceding diagnoses, the process is known as a second-order Markov chain, and so on.

In general, a first-order Markov chain for, for example, depression, can be defined in terms of the proportion of depressed subjects at the beginning of the study, and the two transition probabilities, P_{WD} and P_{DD}. If one considers, instead of probabilities, the estimated odds on being depressed; that is

$$\text{odds}_{WD} = \frac{n_{WD}}{n_{WW}} \tag{11.1a}$$

and, similarly,

$$\text{odds}_{DD} = \frac{n_{DD}}{n_{DW}} \tag{11.1b}$$

Then the two independent transition probabilities can be replaced by their corresponding odds, and these can be modelled in terms of a simple linear-logistic model. A first-order Markov chain is characterised by stationary odds and can be illustrated graphically by two horizontal lines of points.

Algebraically, the model can be described by

$$\ln(\text{odds}) = \beta_0 + \text{Dep}(j) \tag{11.2}$$

where Dep(j) is the effect of the previous year's diagnosis ($j = 1$ if well, 2 if depressed). In the special case where Dep(j) is zero, consecutive diagnoses are independent, and the pattern of transitions between diagnoses would be characterised by a *Bernoulli process*.

Simple Markov chains have rarely been found to be sufficient to describe social and behavioural processes, although they are usually considered to be a good starting point in an analysis. How might we improve on these models? One way is to consider the possibility that the structure of the time sequences is different for different groups of people. For example, in the depression study we might introduce differences between males and females and also between different age groups. Dropping the explicit reference to the constant term (β_0) and the level indicators for factors Dep, Sex and Age, one possible model would now be

$$\ln(\text{odds}) = \text{Dep} + \text{Sex} + \text{Age} + \text{Sex.Age} \tag{11.3}$$

Such a model implies that age and sex have the same effect on both the transition odds of interest. If they both have different effects, the following would be more appropriate

$$\ln(\text{odds}) = \text{Dep} + \text{Sex} + \text{Age} + \text{Sex.Age} + \text{Dep.Sex} + \text{Dep.Age} \tag{11.4}$$

where the meaning of the parameters should be clear from the context. A further possible addition to the simple first-order Markov chain model would be to allow for an increase in order; that is, to allow for more 'memory' in the model. Second-order chains would be described by a model such as

$$\ln(\text{odds}) = \text{Dep}_{-1} + \text{Dep}_{-2} \tag{11.5}$$

or more realistically,

$$\ln(\text{odds}) = \text{Dep}_{-1} + \text{Dep}_{-2} + \text{Dep}_{-1}.\text{Dep}_{-2} \tag{11.6}$$

where Dep_{-1} and Dep_{-2} are indicating the effects of the subject's state at times $i-1$ and $i-2$, respectively.

A more serious problem with the Markov chain as a model of change is that it is a model involving discrete time points, whereas the data may have been generated by grouping information from state changes that can occur at any time. The choice of the above time points to describe the transition probabilities is entirely arbitrary. Records of depression, for example, may be called at monthly intervals of once every five years. Logistic models such as (11.4), however, may still be a pragmatically useful way of describing patterns of data and form a useful framework within which one can test precise hypotheses.

In the last chapter we discussed survival models for grouped data (see Section 10.4). Data such as those is Table 11.2 can be modelled in the same way but with the recognition that we are now modelling repeated events. Consider the ln(odds) of being depressed in year t. One possible model is

$$\ln(\text{odds}) = \text{Year} + \text{Sex} + \text{Age} + \text{Dep} \tag{11.7}$$

This is analogous to equation (10.18) except that it allows for variability in the state at year $i-1$. It is obviously no longer a model for 'first' episodes of depression. Interactions between any of the terms in (11.7) could also be added to this model. Another possibility is to allow for the number of years since the last positive record of depression. One could also let the probability of the nth event (a year with one or more episodes of depression) be dependent on $(n-1)$, the number of previous events. One could, of course, specify much more complex kinds of dependence on previous history but, in most cases, this should not be necessary. The reader is referred to Allison (1984) and to Dunn and Skuse (1981) for further details.

Now consider a possible model for serial dependencies between repeated quantitative measurements. Let $e_1, e_2, \ldots, e_T$ be uncorrelated random variables, where e_t has a mean of zero and variance σ_t^2 for $t = 1, 2, \ldots, T$. Now define a series of measurements $Y_1, Y_2, \ldots, Y_T$ by

$$Y = e_1$$
$$Y_t = \gamma_t Y_{t-1} + e_t, \quad t = 2, 3, \ldots, T \tag{11.8}$$

If the γ_ts are all equal ($\gamma_t = \gamma$ for all t) and so are the variance terms ($\sigma_t^2 = \sigma^2$ for all t) then (11.8) describes a stationary first-order autoregressive process analogous to the Markov chain described above. If, however, these parameters are permitted to vary with time then (11.8) describes first-order *antedependence* (Gabriel, 1962; Byrne and Arnold, 1983). A measurement at time t (that is, Y_t) is dependent on the value of Y_{t-1} but conditional on the value of Y_{t-1}, it is independent of all previous measurements. In general the set of ordered measurements $Y_1, Y_2, \ldots, Y_T$ is said to have an antedependence structure of order r if the measurement at time t (with t greater than r), given the preceding r measurements, is independent of all further preceding measurements. Kenward (1987) describes a method of assessing the order of a sequence of observations and the reader is referred to this paper for further details. The next section will, however, illustrate the analysis of a data set based on the assumption of first-order antedependence.

11.3   Analysis of Covariance

The two main reasons for carrying out an analysis of covariance are (a) to increase the precision in a randomised experiment, and (b) to reduce bias in an observational study where, because of the lack of randomisation, there might be differences between the observed groups on one or more important prognostic factors. In the latter case the aim is to estimate any group differences in the outcome variable having adjusted for any differences in one or more covariates. In a simple repeated measures experiment or observational study in which a subject is assessed twice — once at the beginning of the study and again at the end — analysis of covariance is one of the obvious possibilities for its analysis (another being an analysis of simple change scores). Here the outcome measure (Y_2, say) is the dependent variable, and the initial measurement (Y_1, say) is the corresponding covariate. If we have a factor Group(i) indicating the effect of being in the ith group then the required linear model is

$$Y_2 = \beta_0 + \text{Group}(i) + \beta_1 Y_1 \tag{11.9}$$

In addition to the assumptions normally made for an analysis of variance, it is assumed that there is no covariate by group interaction (that is, the effect of the covariate is the same within each of the groups). If the effect of the

Table 11.6 Measurements of salsinol (mmol) and dependence status for 14 individuals (reproduced with permission from Hand and Taylor, 1987)

Patient	Group	Day 1	Day 2	Day 3	Day 4
1	2	0·64	0·70	1·00	1·40
2	1	0·33	0·70	2·33	3·20
3	2	0·73	1·85	3·60	2·60
4	2	0·70	4·20	7·30	5·40
5	2	0·40	1·60	1·40	7·10
6	2	2·60	1·30	0·70	0·70
7	2	7·80	1·20	2·60	1·80
8	1	5·30	0·90	1·80	0·70
9	1	2·50	2·10	1·12	1·01
10	2	1·90	1·30	4·40	2·80
11	1	0·98	0·32	3·91	0·66
12	1	0·39	0·69	0·73	2·45
13	1	0·31	6·34	0·63	3·86
14	2	0·50	0·40	1·10	8·10

Group 1: moderately dependent
Group 2. severely dependent

covariate is not, in fact, linear then model (11.9) can easily be modified to allow for this nonlinearity. Equation (11.9) is simply a regression model in which the group effects are assessed after fitting the effect of the covariate, Y_1.

Consider for example, the data given in Table 11.6. These data were collected during an investigation into the role that the alkaloid salsinol plays in bodily dependence on alcohol. Fourteen individuals attending an alcoholism treatment unit were observed over a period of four days immediately after being admitted to the unit, measurement of salsinol being made from urine samples taken daily throughout the study period. The individuals were categorised as being in one of two groups: those considered to be severely dependent and those judged to be only moderately dependent. The response variables in this study are the four repeated measurements of urine concentrations of salsinol. As the inspection of the distributions of the measurements indicates skewness a log transformation is carried out prior to any analysis. The logged measurements for the four days are labelled Y_1 to Y_4, respectively.

First we will analyse the data as if Y_2 and Y_3 were not available. The results of an analysis of covariance, using Y_4 as the response and Y_1 as the covariate, are given in Table 11.7. A plot of Y_4 against Y_1 for each of the two groups is given in Figure 11.1. It will be left to the reader to interpret these results.

Now return to the full data set. Kenward (1987) has described the use of analysis of covariance for the comparison of the complete profiles under the assumption of antedependence. Here we will make the (possibly unrealistic) assumption that the four measurements of salsinol concentration display first-order antedependence. Because of the limited nature of the data there would not be a lot to be gained from a formal analysis of the antedependence structure. Here the first-order assumption is made simply to illustrate Kenward's method. First we carry out a simple analysis of variance, with Y_1 as the dependent variable, and extract the group sum of squares (h_1) and the corresponding

Table 11.7 Salsinol data: analysis of covariance with Y_4 as the dependent variable

Source	d.f.	S.S.	M.S.	F	P
Y_1	1	3·674	3·674	10·266	0·008
Group	1	1·744	1·744	4·873	0·049
Residual	11	3·936	0·358		
Total	13	9·354			

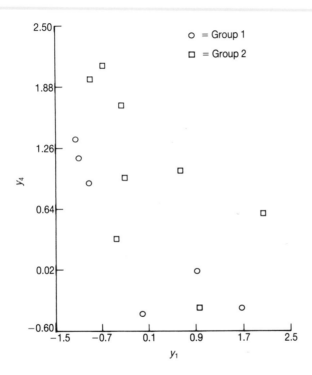

Figure 11.1

residual sum of squares (d_1). We next carry out an analysis of covariance using Y_2 as the dependent variable and Y_1 as the covariate. Again we extract the group sum of squares (h_2) and the corresponding residual sum of squares (d_2). The analysis of covariance is repeated for Y_3 and Y_4 as dependent variables where, for Y_3, Y_2 is the covariate and for Y_4, Y_3 is the covariate. In each case the group sum of squares and the residual sum of squares are extracted. The results are summarised in Table 11.8. Crowder and Hand (1990) illustrate the use of a series of covariance analysis corresponding to the assumption of second-order antedependence. If, for the ith step, the group sum of squares is h_i, and the corresponding residual sum of squares is d_i, then the overall test

Table 11.8 Salsinol data: derivation of Kenward's statistic to compare profiles

Step (i)	Residual S.S. (d_i)	Group S.S. (h_i)	$\ln\dfrac{d_i}{d_i+h_i}$	$\phi(m)$
1	14·220	0·174	−0·0122	0·0871
2	8·766	0·077	−0·0087	0·0955
3	7·259	0·459	−0·0613	0·0955
4	8·140	1·138	−0·1309	0·0955
		Sum	−0·2131	0·3735

$$X^2 = \frac{4(0·2131)}{(0·3735)}$$
$$= 2·282 \text{ with 4 d.f.}$$

statistic to test the difference between the profiles of the four measurements is given by

$$X^2 = \frac{-p\sum_{i=1}^{p}\ln\left(\frac{d_i}{d_i+h_i}\right)}{\sum_{i=1}^{p}\phi(m)} \tag{11.10}$$

where p is the number of repeated measurements (here $p = 4$) and

$$\phi(m) = \frac{(2m-1)}{2m(m-1)} \tag{11.11}$$

where

$$m = n_i - q - r_i \tag{11.12}$$

In the latter expression n_i is the number of observations at the ith step (here $n_i = 14$ for all i), q is the rank of the design matrix (here $q = 2$), and r_i is the minimum of $i - 1$ or r (the order of the outer dependence structure (here $r_i = 0$ when $i = 1$ and $r_i = 1$ for $i = 2, 3$ or 4). Kenward (1987) shows that X^2 in (11.10) has an asympotic chi-square distribution under the null hypothesis of no group differences. There does not appear to be a significant difference between the two salsinol profiles.

One attraction of Kenward's approach is that the method easily copes with dropouts — provided, of course, that the reason for the subject leaving the study is unrelated to the measurements being taken. Here one simply bases each step of the analysis on the measurements actually available and adjusts the value of m in (11.12) accordingly.

11.4   Mixed and Random Effects Models

Table 11.9 presents the results of an experiment to assess the precision of a simple planimeter. Each of the map distances is measured five times, yielding a table of 50 measurements. Let the jth measurement on the ith route be represented by X_{ij} ($i = 1, 2, \ldots, 10; j = 1, 2, \ldots, 5$). Note that there is no correspondence between the jth measurement on any one route with the jth measurement on another. The repeated measurements are nested within routes and a better representation of the measurements might be $X_{j(i)}$ to indicate the

Table 11.9 Measurements of map distances (arbitrary units) using a planimeter (from Dunn, 1989)

Route	Replicate measurements				
1	48·0	47·5	48·0	49·5	48·0
2	32·0	32·5	33·5	31·5	32·5
3	12·5	14·0	14·0	14·5	14·5
4	29·5	30·0	29·0	29·0	30·0
5	50·0	50·0	49·5	49·5	50·5
6	55·5	55·5	56·0	56·0	56·5
7	22·0	23·5	23·0	24·0	22·0
8	50·5	50·0	51·5	51·0	50·0
9	26·5	26·0	26·0	25·5	25·5
10	24·0	24·0	24·5	23·0	23·5

jth replication within route i. Where there is no risk of confusion, however, the simpler terminology will be used.

Now a suitable specification of a measurement model for the map distances in Table 11.9 is the following:

$$X_{ij} = \mu + \tau_i + \epsilon_{ij} \qquad (11.13)$$

Here μ is the overall mean, τ_i is the departure from this mean which is characteristic of route i, and ϵ_{ij} can be considered to be the measurement error for the jth measurement on route i. Unlike models described earlier in Chapters 7 and 8, this model contains two random components: τ_i and ϵ_{ij}. The τ_i are not fixed. If we were to take another random sample of map routes we would obviously get another characteristic set of τ_i. We assume that τ_i is a random variable with a mean of zero and variance σ_τ^2. Similarly, ϵ_{ij} is another random variable with mean zero and variance σ_ϵ^2. It is assumed that the errors of measurement are uncorrelated with each other and with other components of the model. Similarly, the τ_i are also assumed to be uncorrelated. From these assumptions it follows that

$$\text{Var}(X_{ij}) = \sigma_\tau^2 + \sigma_\epsilon^2 \qquad (11.14)$$

$$\text{Cov}(X_{ij}, X_{ih}) = \sigma_\tau^2 \quad (h \neq j) \qquad (11.15)$$

and

$$\text{Cov}(X_{ij}, X_{kj}) = 0 \quad (i \neq k) \qquad (11.16)$$

Equation (11.13) and the corresponding assumptions describe a random effects model. In the analysis of the data set such as that in Table 11.9, the primary interest lies in the estimation of the two variance components, σ_τ^2 and σ_ϵ^2, and, in particular, the ratio $\frac{\sigma_\tau^2}{\sigma_\tau^2 + \sigma_\epsilon^2}$. This ratio is defined to be the *intraclass correlation*. Details of the estimation of variance components will not be given here, and the reader is referred to Chapter 6 of Dunn (1989).

Returning to the salsinol data, a possible linear model to describe them is the following:

$$X_{ijk} = \mu + \alpha_i + \pi_k + \gamma_{ik} + \omega_{ij} + \epsilon_{ijk} \qquad (11.17)$$

Table 11.10 Analysis of variance for logged salsinol measurements

Source of variation	d.f.	M.S.	F	P
Between groups	1	1·51	2·48	0·14
Subjects within groups	12	0·61		
Between times	3	1·54	1·79	0·17
Groups by times	3	0·13	0·16	0·92
Subjects within groups by times	36	0·86		

where X_{ijk} represents the logarithm of salsinol measurement for the jth subject in the ith group on the kth day. Note here that subjects are nested within groups. The terms on the right hand side of (11.17) correspond to the following effects: μ is the overall mean; α_i is the effect of being in the ith group ($i = 1, 2$); π_k is the effect of the kth day ($k = 1$ to 4); γ_{ik} is the group by day interaction; ω_{ij} is the effect of the jth subject in the ith group; and ϵ_{ijk} is the residual error term. Here, the terms α_i, π_k and γ_{ik} are fixed effects, and the terms ω_{ij} and ϵ_{ijk} are random variables or random effects. This model is an example of a *mixed model*. Further details of models of this type can be found in Winer (1971) and in Montezun, Blouin and Malone (1984). If we assume that the two random effects are uncorrelated then

$$\text{Var}(X_{ijk}) = \sigma_\omega^2 + \sigma_\epsilon^2 \qquad (11.18)$$

and

$$\text{Cov}(X_{ijk}, X_{ijl}) = \sigma_\omega^2, \quad k \neq l \qquad (11.19)$$

where σ_ω^2 is the variance of the ω_{ij} and σ_ϵ^2 the variance of the ϵ_{ijh}. Again, the intraclass correlation is defined by

$$\rho = \frac{\sigma_\omega^2}{\sigma_\omega^2 + \sigma_\epsilon^2} \qquad (11.20)$$

In summary, the implications of this mixed model are that (a) the variances of the observations are the same at each time point, and (b) the correlations between observations at any two different times are the same for every choice of times. The covariance matrix of the measurements displays what is known as *compound symmetry*.

If we are now prepared to make the further assumption that the logged salsinol measurements are normally distributed (remembering that the model is being applied to the logarithms of the measurements) then the results can be analysed through the use of an analysis of variance (see Winer, 1971 and Monlezun *et al.*, 1984 for details of how to construct appropriate F-statistics). The results are shown in Table 11.10. If the compound symmetry assumption does not appear to be justified for any particular data set then an analysis such as that given in Table 11.10 can be still carried out but the F-tests are modified using a correction suggested by Greenhouse and Geisser (1959). Details are given in Everitt (1989). Finally, tests of the equality of covariance matrices for the two groups can be carried out using the method suggested by Box(1950).

Now suppose that the effect of time can be represented by a linear trend. The model in (11.17) might now be replaced by

$$X_{ijk} = \beta_0 + \alpha_i + \omega_{ij} + (\beta_j + \beta_{ij})t_k + \epsilon_{ijk} \qquad (11.21)$$

The variable t_k is the observed time of the kth measurement. Here the fixed effects are the intercept term, β_0, α_i, the effect of the ith group on the intercept ($i = 1, 2$) and β_i, the linear effect of time in the ith group. The random effects are ω_{ij}, the effect on the intercept of subject j within group i, β_{ij}, the variation in linear effect of time that is characteristic of subject j within group i, and the residual term, ϵ_{ijk}. We assume that the means of the random effects are all zero and that they are uncorrelated with and across individuals. The consequences of this model are

$$\mathrm{Var}(X_{ijk}) = (\sigma_\omega^2 + 2t_k\sigma_{\omega\beta} + t_k^2\sigma_\beta^2) + \sigma_\epsilon^2 \qquad (11.22)$$

$$\mathrm{Cov}(X_{ijk}, X_{ijl}) = \sigma_\omega^2 + (t_k + t_l)\sigma_{\omega\beta} + t_jt_l\sigma_\beta^2, \quad j \neq l \qquad (11.23)$$

where $\sigma_{\omega\beta}$ is the covariance of ω_{ij} and β_{ij}, $\sigma_\omega^2 = \mathrm{Var}(\omega_{ij})$ and $\sigma_\epsilon^2 = \mathrm{Var}(\epsilon_{ijk})$. Model (11.21) could be further extended by the introduction of quadratic and possibly cubic time trends. Methods of fitting models such as (11.21) are described in Crowder and Hand (1990), Goldstein (1987), and Laird and Ware (1982). Stanek (1990) will also be of interest. Here we will not concentrate on fitting trend components explicitly as in (11.2) but will follow the more pragmatic approach advocated by Matthews *et al.* (1990). Basically, this involves the extraction of the features of interest for each individual separately followed by statistical analysis of these summary measures by univariate methods (in the following Section) or through the use of multivariate tests (section 11.6).

11.5   Extraction of Response Features and Trend Components

Figure 11.2 illustrates the idea of basing an analysis on summary statistics for each individual subject. These summary statistics are chosen to reflect a particular feature of the response in these serial measurements. Each individual in an experiment or observational study could be assumed to generate a series of serial measurements similar to that in Figure 11.2 but the pattern might differ in certain ways. The obvious summary statistics for Figure 11.2 are (a) the maximum response, (b) the time taken to achieve the maximum response, and (c) the total response (measured as the overall mean response for equally spaced measurements or area under the curve for irregularly spaced measurements). Each of these individual-based summary statistics is then entered into the second stage of the analysis to illustrate group differences, and so on. If the summary statistics for the individuals have corresponding estimates of precision (being based on different numbers of measurements, for example), then the investigator might wish to use a weighted analysis. Matthews *et al.* (1990) also point out that the subject-based summary measures have considerable advantages for plotting because the usual graphical methods such as box-plots and scatter diagrams (together with those described in Chapter 3) can be applied to them.

Returning to the salsinol data (Table 11.6) what would be the appropriate summaries here? The obvious ones are the overall mean of the four logged values and a simple measure of trend. This could be obtained through the

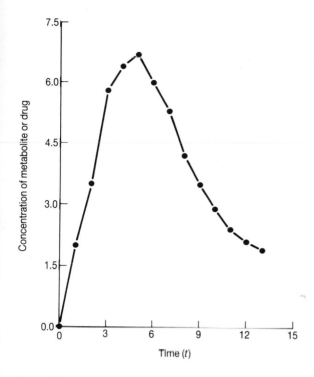

Figure 11.2

use of a simple linear regression of the logarithm of salsinol concentration on time. If Y_k is the logarithm of the kth measurement ($k = 1$ to 4) then

$$y_k = \beta_0 + \beta_1 t_k \tag{11.24a}$$

Each subject here provides a separate estimate of β_1 and these estimates could, for example, be used in a simple t-test for the comparison of the two severity groups. One could also fit quadratic and cubic components of trend to the data for each subject. Here

$$y_k = \beta_0 + \beta_1 t_k + \beta_2 t_k^2 + \beta_3 t_k^3 \tag{11.24b}$$

Differences between the two groups in β_1, β_2 and β_3 could now be estimated or tested. A more common and generally more useful approach is to extract orthogonal polynomial trends or contrasts (see Hand and Taylor, 1987 for a discussion of these and other possible contrasts). The linear, quadratic and cubic trends represented by T_L, T_Q and T_C, respectively, are calculated from

$$T_L = -3Y_1 - Y_2 + Y_3 + 3Y_4 \tag{11.25}$$

$$T_Q = Y_1 - Y_2 - Y_3 + Y_4 \tag{11.26}$$

$$T_C = -Y_1 + 3Y_2 - 3Y_3 + Y_4 \tag{11.27}$$

Table 11.11 Orthogonal trend components from logged salsinol data

Patient	Group	Mean	T_L	T_Q	T_C
1	2	−0·12	2·70	0·25	−0·29
2	1	0·14	8·02	−0·43	−1·34
3	2	0·63	4·48	−1·26	−0·73
4	2	1·19	6·68	−2·09	0·38
5	2	0·46	8·50	0·24	3·28
6	2	0·13	−4·56	0·69	0·54
7	2	0·94	−3·63	1·50	−3·79
8	1	0·45	−5·38	0·83	−4·10
9	1	0·45	−3·35	0·07	0·98
10	2	0·85	2·38	−0·07	−3·27
11	1	−0·05	1·32	−0·66	−7·90
12	1	−0·18	5·57	0·64	1·67
13	1	0·39	5·26	−1·21	9·45
14	2	0·14	9·37	2·22	−0·25

These, of course, can be augmented by the overall total (or mean) so that the four derived measures provide a complete description of the original data. The absolute values of the coefficients in equations (11.25) to (11.27) are completely arbitrary and they are often scaled so that the sum of their squared values equals unity. Here we would have

$$T'_L = 0.671 Y_1 - 0.224 Y_2 + 0.224 Y_3 + 0.671 Y_4 \qquad (11.28)$$

$$T'_Q = 0.500 Y_1 - 0.500 Y_2 - 0.500 Y_3 + 0.500 Y_4 \qquad (11.29)$$

$$T'_C = -0.224 Y_1 + 0.671 Y_2 - 0.671 Y_3 + 0.224 Y_4 \qquad (11.30)$$

together with

$$M = 0.500 Y_1 + 0.500 Y_2 + 0.500 Y_3 + 0.500 Y_4 \qquad (11.31)$$

T'_L, T'_Q and T'_C provide a set of *orthonormal contrasts*. For the salsinol data, the orthogonal trend components defined by (11.25) to (11.27) are given in Table 11.11. Univariate t-tests or analyses of variance can now be carried out separately for each of the four summary measures to assess (a) whether there are differences between groups, and (b) whether each trend component is significantly different from zero. This will be left as an exercise for the reader. The usual approach is to look at each trend in turn, beginning with the most complex (here the cubic brand) and when one produces a significant result the sequence of tests is interrupted and no further trends are examined. One might, of course, wish to test group differences for the three trend measures taken together. This is the subject of the next section.

11.6   Multivariate Tests

First consider the comparison of two groups. For a univariate measurement x with mean values of $\bar{x}_1$, and $\bar{x}_2$ in groups 1 and 2, respectively, we can define a *t*-statistic given by

$$t = \frac{\bar{x}_1 - \bar{x}_2}{\sqrt{(\frac{1}{n_1} + \frac{1}{n_2})s^2}}$$

(11.32)

where s^2 is the pooled estimate for the common within-group variance and n_1 and n_2 are the respective sample sizes for group 1 and group 2. This statistic can be compared with a *t*-variate with $(n_1 + n_2 - 2)$ d.f. This *t*-statistic can also be squared and compared to the corresponding Fisher's *F* with 1 and $(n_1 + n_2 - 2)$ d.f. That is

$$t^2 = \frac{(\bar{x}_1 - \bar{x}_2)^2}{(\frac{1}{n_1} + \frac{1}{n_2})s^2}$$

(11.33)

or, alternatively

$$t^2 = \frac{n_1 n_2}{(n_1 + n_2)} \frac{(\bar{x}_1 + \bar{x}_2)^2}{s^2}$$

(11.34)

If we now move to vectors of measurements with means represented by $\bar{\mathbf{x}}_1$ and $\bar{\mathbf{x}}_2$ for groups 1 and 2, respectively, then the multivariate test statistic analogous to that defined by (11.34) is given by

$$T^2 = \frac{n_1 n_2}{(n_1 + n_2)}(\bar{\mathbf{x}}_1 - \bar{\mathbf{x}}_2)'\mathbf{S}^{-1}(\bar{\mathbf{x}}_1 - \bar{\mathbf{x}}_2)$$

(11.35)

where **S** is the pooled estimate of the within-groups variance — covariance matrix. The statistic defined by (11.35) is called Hotelling's T^2. It can be shown that

$$F = \frac{(n_1 + n_2 - p - 1)}{(n_1 + n_2 - 2)p} T^2$$

(11.36)

is distributed as an *F*-variate with *p* and $n_1 + n_2 - p - 1$ d.f. under the null hypothesis of no group differences. Here *p* is the number of variable in the vector **x**. For the logged salsinol values $T^2 = 5.87$ and the corresponding *F* is 1.10 with 4 and 9 d.f. ($P \leq 0.413$).

In a univariate analysis of variance we are essentially concerned with test statistics which are proportional to the ratio of a hypothesis sum of squares (*H*) to the error or residual sum of squares (*E*). For a hypothesis with *g* d.f. and a residual sum of squares with *q* d.f. the required F-statistic is

$$F = \frac{qH}{gE}$$

(11.37)

with *g* and *q* d.f. In the multivariate case we partition the overall sums of squares and cross-products matrix (**T**) into a hypothesis sums of squares and cross-products matrix (**H**) and a corresponding error matrix (**E**), so that

$$\mathbf{T} = \mathbf{H} + \mathbf{E}$$

(11.38)

The multivariate analogue of the ratio of the hypothesis sum of squares to the residual sum of squares is provided by $\mathbf{HE}^{-1}$. Let the eigenvalues of $\mathbf{HE}^{-1}$ be $\lambda_1, \lambda_2, \ldots$, in decreasing order of size. One commonly used test statistic is *Wilk's lambda*, defined by

$$\Lambda = \sum_j 1/(1 + \lambda_j)$$

(11.39)

Others are given in Hand and Taylor (1987). Wilk's lambda is in fact, the likelihood ratio statistic for testing that a hypothesis comparison is zero, assuming multivariate normality and further assuming equality of covariance matrices across groups. In general a transformation of Wilk's lambda can be derived which is approximately distributed as an F-variate with the appropriate degrees of freedom (see, for example, Chatfield and Collins, 1980). When there are two groups, however,

$$T^2 = (n_1 + n_2 - 2)\frac{1 - \Lambda}{\Lambda} \tag{11.40}$$

This equality, of course, leads to the F-statistic in (11.36). For the four logged salsinol measures, $\Lambda = 0.671$.

Returning to the logged salsinol measurements yet again, how do we apply the above multivariate tests to test for group differences? The Hotelling T^2-statistic given above is not particularly useful if we wish to look specifically at trend differences. A much more sensible strategy would be to carry out a univariate t-test for the overall mean (this will be left as an exercise for the reader). We then carry out a multivariate test for the three orthogonal trend components. If this trend test is significant we then proceed to the univariate tests described in the last section. For the three contrasts together the T^2-statistic is 0.71 and $\Lambda = 0.944$. The corresponding F-statistic is 0.197 with 3 and 10 dif. ($P \leq 0.896$).

Finally in this section, we illustrate the use of Hotelling's T^2 to compare two profiles that show first-order antedependence. This is a simpler alternative to the method of Kenward (1987) which was described in Section 11.3. It is, however, dependent on the fact there are no drop-outs or other missing values. Details of the derivation of the test statistic are given in Byrne and Arnold (1983). We first compute T^2-statistics for pairs of measurements $(Y_1, Y_2), (Y_2, Y_3), \cdots, (Y_{p-1}, Y_p)$. For the logged salsinol values there are three T^2-statistics to be computed and they are equal to 0.254, 0.859 and 2.548, respectively. We then compute the univariate t-statistics for each time point leaving out the first and the last time points. For the salsinol data they are 0.094 and 0.763. The Byrne and Arnold statistic is given by

$$F = \sum T^2 - \sum t^2 \tag{11.41}$$

and, under the null hypothesis of no group differences, this is approximately distributed as an F-variate with p and $n_1 + n_2 - 2$ d.f. Here F is 2.80.

11.7   Trend Analysis for Categorical Responses

Table 11.3 illustrates on experiment where each subject contributes a profile of three binary responses. One approach to the analysis of these data is through the use of log-linear models although this is not ideal for our present purposes. Here treatment method and diagnosis are the explanatory factors and the profile of binary responses can be considered as a structured multinomial response (see Section 9.4). If there were no association between the explanatory factors and outcome the fitted model would be

$$\text{Treat} * \text{Diagnosis} + T1 * T2 * T3 \tag{11.42}$$

Table 11.12 Parameter estimates for the log-linear
model of the Koch data (Table 11.3)

Parameter	Estimate	Standard error
β_0	−1·47	0·31
Treat(2)	−2·58	0·42
Diag(2)	1·96	0·30
T1(2)	0·08	0·37
T2(2)	0·36	0·33
T3(2)	1·14	0·33
Treat(2).Diag(2)	0·70	0·28
T1(2).T2(2)	0·29	0·45
T1(2).T3(2)	−0·30	0·41
T2(2).T3(2)	−0·32	0·33
T1(2).T2(2).T3(2)	0·15	0·52
Diag(2).T1(2)	−1·44	0·26
Diag(2).T2(2)	−1·23	0·26
Diag(2).T3(2)	−1·22	0·31
Treat(2).T1(2)	−0·07	0·27
Treat(2).T2(2)	1·05	0·26
Treat(2).T3(2)	2·07	0·31

where T1, T2 and T3 are each two-level factors which are indicators of the
binary response at the three times (1=abnormal, 2=normal, in each case),
respectively. If we now let diagnosis influence each of the outcomes we change
to

$$\text{Treat} * \text{Diagnosis} + \text{T1} * \text{T2} * \text{T3} + \text{Diagnosis} * \text{T1}$$

$$+ \text{Diagnosis} * \text{T2} + \text{Diagnosis} * \text{T3} \tag{11.43}$$

The deviance for this model is 86.20 with 18 d.f. This can be regarded as
the starting point for a test of the association between treatment method and
outcome. We now fit

$$(11.43) + \text{Treat.T1} + \text{Treat.T2} + \text{Treat.T3} \tag{11.44}$$

giving a deviance of 15.43 with 15 d.f. Estimates are displayed in Table 11.12.
The only estimates of real interest in the present context are those for Treat
(2).T1(2), Treat(2).T2(2) and Treat(2).T3(2). The interpretation of these estimates is fairly clear-cut: the effectiveness of the new treatment, when compared
to the standard, is increasing with time. There does not appear to be any trend
in the diagnosis by time interactions. Nor does there appear to be much
association between outcomes at the three different time points.

Now consider the outcome of the trial reported in Table 11.3 from a different
perspective. First we create a variate t, with scores 0, 1 and 2 at weeks 1, 2
and 3 respectively. Koch *et al.* (1977) justified this scaling by noting that if
the time measurement reflects cumulative dosage of the drugs, then a logistic
response should be a linear function of the logarithm of time. Let Treat now
be a dummy variable which is 0 for the standard treatment and 1 for the new
one. Similarly let Diag be 0 for mild symptoms and 1 for severe. We now
postulate a model for the odds of being normal at the three time points which
is essentially a model for the marginal distributions of a binary response (see

Koch *et al.*, 1977; Agresti, 1990). This model is

$$\ln(\text{odds}) = \beta_0 + \beta_1 \text{Diag} + \beta_2 \text{Treat} + \beta_3 t \tag{11.45}$$

   This is compared with

$$\ln(\text{odds}) = \beta_0 + \beta_1 \text{Diag} + \beta_2 \text{Treat} + \beta_3 t + \beta_4 t.\text{Treat} \tag{11.46}$$

The new parameter, β_4, is a measure of the treatment by time interaction. Care should be taken in fitting these two models since there is a need to allow for association between the outcomes at the three follow-up times. Despite their superficial similarity to earlier logistic models they cannot easily be fitted by maximum likelihood methods. They are usually fitted using weighted least squares, the weights being determined from the variance and covariances characteristic of the multinomial distribution. That is, each row of Table 11.3 is regarded as an independent multinomial response and, if the observed proportion of the subjects that fall into column j of the row is P_j then the estimate of the variance of that estimate is given by $\frac{P_j(1-P_j)}{n}$, where n is the total count for the row. Similarly, the covariance of the proportions P_j and P_k in columns j and k of a given row is estimated by $\frac{-P_j P_k}{n}$. Details are given in Koch *et al.* (1977). Agresti (1990) reports a chi-square statistic of 31.1 with 8 d.f. for model (11.45) and a chi-square of 4.2 with 7 d.f. for model (11.46). Clearly there is a highly significant treatment by time interaction. His estimate for β_4 is 0.99 (s.e. 0.19) and the estimate for β_1 is 1.28 (s.e. 0.15). These are almost exactly what one would expect from an examination of the parameter estimates given in Table 11.11 and the interpretation is, of course, the same. The advantage of the models for the marginal responses as in (11.45) and (11.46) is their relative simplicity and ease of interpretation. This advantage becomes overwhelming when one considers models for ordinal responses measured at repeated intervals (see Agresti, 1989 and 1990).

Exercises

11.1 Reanalyse the logged salsinol data using Kenward's method under the assumption of first-order antedependence. Here, however, analyse the data as if subject 2 dropped out of the experiment after day 2, subject 5 dropped out after day 1 and subject 9 had missing data for day 4.

11.2 Carry out a univariate analysis of variance of the logged salsinol data modifying the F-tests through the use of the Greenhouse and Geisser correction.

11.3 Carry out both univariate and multivariate tests of group differences for the orthogonal components of trend given in Table 11.11. Through the use of your computer output from the multivariate analysis, check that the relationships between T^2, Wilk's Λ and the corresponding F-statistic are as we described in Section 11.6.

11.4 Table 11.13 summarises the transitions for the sequence of behaviours recorded in Table 11.1. Analyse the two-way contingency table through the use of a saturated log-linear model and compare it with the corresponding main effects model. Next, treating the state at time t as an

Table 11.12   First-order transitions for the autistic child's behavioural sequence (from Dunn and Clark, 1982)

(a) Frequencies			State at time *t*			
			1	2	3	Total
		1	60	22	4	86
State at time *t*−1		2	21	100	11	132
		3	4	11	6	21
	Total		85	133	21	

(b) Probabilities			State at time *t*			
			1	2	3	Total
		1	0·70	0·26	0·04	1
State at time *t*−1		2	0·16	0·76	0·08	1
		3	0·19	0·52	0·29	1

ordinal response use a logistic model to investigate the influence of the state at $t-1$ on this ordered response (see Agresti, 1989 and 1990).

11.5 Repeat the analysis of the data in Table 11.3 using the log-linear models described by equations (11.43) and (11.44). Investigate the effects of dropping (a) the three-way interaction $T1.T2.T3$, (b) the three way interaction *and* $T1.T3$, and (c) all of the mutual interactions between $T1$, $T2$ and $T3$. Interpret the results.

12

Discriminant Analysis

12.1 Introduction

The discriminant analysis situation arises when one has two types of multi-variate observations; the first, called *training samples*, are those whose group identity (i.e. membership in a specific one of g given groups), is known *a priori*, and the second, referred to as *test samples*, consists of observations for which group membership is unknown, and which have to be assigned to one of the g groups. Areas where such a problem might be of interest are numerous; some examples are

(1) *Medical diagnosis.* Here the variables describing each individual might be the results of various clinical tests and the groups could be collections of patients known to have different diseases.

(2) *Archaeology.* Here the variables might be particular measurements on the artefacts of interest and the groups an existing taxonomy of such objects.

(3) *Speech recognition.* Here the objects to be classified are usually wave-forms and the variables a set of acoustical parameters extracted from the utterance of a specific word by an individual.

(Many other examples of where the discrimination problem is of interest are given in Hand, 1981b).

The discriminant analysis situation differs from that where cluster analysis techniques (see Chapter 6) would be helpful, because of the existence of a set of relevant groups which are known *a priori*. But although discriminant analysis and cluster analysis constitute a useful dichotomy of classification problems, many real-life problems involve features of both. For example, Hosmer(1973) describes a situation in which an investigator wishes to determine the mean and variance of length of male and female halibut. Unfortunately the sex of halibut can be determined by humans only by dissection of the fish. Small samples taken from research vessels are sexed in this way, but much larger samples are available from commercial catches which cannot be sexed. Far more accurate estimates of the parameters of interest might be obtained if the

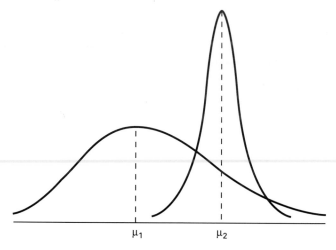

Figure 12.1 Two normal densities with different means and variances

labelled and unlabelled samples could be combined in some way, and Hosmer describes various procedures which might be helpful.

The most commonly used methods of discriminant analysis are Fisher's *linear discriminant function* and *linear logistic discrimination*, both of which are described in Section 12.3. In the next section however, the discrimination problem is considered in a more general way.

12.2   Discrimination Rules

The discrimination problem may be stated formally as follows:

Assume we have g populations or groups $D_1, D_2, \ldots, D_g$. The object of discriminant analysis is to allocate an individual to one of these groups on the basis of a set of observations, $x_1, x_2, \ldots, x_p$. If associated with each population D_j there is a probability density function for the measurements, of the form $f_j(\mathbf{x})$ where $\mathbf{x}' = [x_1, x_2, \ldots, x_p]$, then an intuitively sensible rule for the allocation process would be as follows.

Allocate the individual with vector of scores, $\mathbf{x}$, to D_j if

$$f_j(\mathbf{x}) = \max_{i=1,\ldots,g} f_i(\mathbf{x}) \tag{12.1}$$

To illustrate this rule let us consider a number of simple examples.

(1)   Suppose we have one binary variable, x, and two groups, D_1 and D_2. In D_1 assume we have

$$Pr(x = 0) = Pr(x = 1) = \frac{1}{2} \tag{12.2}$$

whilst in D_2

$$Pr(x = 0) = \frac{1}{4} \qquad Pr(x = 1) = \frac{3}{4} \tag{12.3}$$

The rule specified in (12.1) will now allocate an individual with $x = 0$ to D_1 and an individual with $x = 1$ to D_2.

(2) Now assume we have a single continuous variable, x, and again two groups. In D_1 the variable has a normal distribution with mean μ_1 and variance σ_1^2, and in D_2 a normal distribution with mean μ_2 and variance σ_2^2. (Assume $\mu_1 > \mu_2$ and $\sigma_1 > \sigma_2$; Figure 12.1 illustrates this situation.) An individual with score x will, according to (12.1), be allocated to D_1 if

$$f_1(x) > f_2(x) \tag{12.4}$$

giving, after a little algebra, the following rule: allocate to D_1 if

$$\frac{\sigma_1}{\sigma_2} \exp\left[-\frac{1}{2}\left[\frac{(x-\mu_1)^2}{\sigma_1^2} - \frac{(x-\mu_2)^2}{\sigma_2^2}\right]\right] > 1 \tag{12.5}$$

On taking logarithms and rearranging, this inequality becomes

$$x^2\left[\frac{1}{\sigma_1^2} - \frac{1}{\sigma_2^2}\right] - 2x\left[\frac{\mu_1}{\sigma_1^2} - \frac{\mu_2}{\sigma_2^2}\right] + \left[\frac{\mu_1^2}{\sigma_1^2} - \frac{\mu_2^2}{\sigma_2^2}\right] < 2\ln\frac{\sigma_1}{\sigma_2} \tag{12.6}$$

It is easy to demonstrate that the set of xs for which this inequality is satisfied falls into two distinct regions, one corresponding to low values of x and the other to high values. Examination of Figure 12.1 indicates that such a rule is clearly sensible.

In the special case where $\sigma_1 = \sigma_2$ it can be shown that $f_1(x)$ exceeds $f_2(x)$ if

$$|x - \mu_2| > |x - \mu_1| \tag{12.7}$$

So if $\mu_1 > \mu_2$ an individual with score x is allocated to D_2 if

$$x > \frac{1}{2}(\mu_1 + \mu_2) \tag{12.8}$$

More interesting, of course, is the multivariate situation, where in one group the vector random variable, $\mathbf{x}$ is assumed to have a multivariate normal density with mean $\boldsymbol{\mu}_1$ and covariance matrix, $\boldsymbol{\Sigma}$ and in the other a multivariate normal density with mean $\boldsymbol{\mu}_2$ and the same covariance matrix, $\boldsymbol{\Sigma}$. The discriminant rule based on (12.1) now leads to a rule, allocate an individual with vector of scores $\mathbf{x}$ to D_1 if

$$\boldsymbol{\alpha}'(\mathbf{x} - \boldsymbol{\mu}) > 0 \tag{12.9}$$

where

$$\boldsymbol{\alpha} = \boldsymbol{\Sigma}^{-1}(\boldsymbol{\mu}_1 - \boldsymbol{\mu}_2) \text{ and } \boldsymbol{\mu} = \frac{1}{2}(\boldsymbol{\mu}_1 + \boldsymbol{\mu}_2) \tag{12.10}$$

In certain situations it might be sensible to assume that the members of some groups are intrinsically more (or less) likely to be observed than members of others. For example, in medicine the common cold is a more likely disease than polio. If the g groups are known to have *prior probabilities* $\Pi_1\Pi_2, \ldots, \Pi_g$, the rule given in (12.1) changes to allocating an individual with vector of scores, $\mathbf{x}$, to the population for which

$$\Pi_j f_j(\mathbf{x}) \tag{12.11}$$

is a maximum.

In the case of two groups described by multivariate normal densities with a common covariance matrix the consideration of prior probabilities changes the rule in (12.1) to

$$\boldsymbol{\alpha}'(\mathbf{x} - \boldsymbol{\mu}) > \ln\frac{\Pi_2}{\Pi_1} \tag{12.12}$$

12.3   Methods of Discriminant Analysis

Several methods for discriminant analysis have been proposed and a comprehensive review is available in Hand (1981b). Differences between methods arise because of the variety of distributional assumptions made about the variables describing each object or individual to be classified. As in other areas of statistics, the methods based on the assumption of normality are the ones most widely used in practice, and so we begin this section with a description of such methods.

12.3.1   Classical Linear Discriminant Analysis

The discrimination rules derived for normally distributed variables in Section 12.2 involve knowledge of the population values of mean vectors and covariance matrices. In practice of course such knowledge is not available and consequently the rules cannot be used in the form given. If however we are willing to assume that our groups are described by multivariate normal densities with different means but the same covariance matrix, then rules such as (12.9) and (12.12) may be used by simply replacing population values, μ_1, μ_2 and Σ with corresponding sample values, $\bar{x}_1, \bar{x}_2$ and S, where S is the pooled within-group sample covariance matrix given by

$$S = \frac{1}{n_1 + n_2}[n_1 S_1 + n_2 S_2] \tag{12.13}$$

where n_1 and n_2 are the sample sizes in the two groups and S_1 and S_2 are the group covariance matrices. Now (12.9) becomes, allocate an individual with vector of scores x to D_1 if

$$a'[x - \frac{1}{2}(\bar{x}_1 + \bar{x}_2)] > 0 \tag{12.14}$$

where

$$a = S^{-1}(x_1 - x_2) \tag{12.15}$$

The function $a'x$ is known as the *linear discriminant function*. As we have seen it arises from assuming multivariate normality within each population; it was however first suggested by Fisher (1936) using an argument not directly involving the assumption of normality. Fisher's idea was to find a linear combination of the p variables which separates the two training samples as much as possible, and he showed that for any such combination, $a'x$, the squared difference between the two sample means, divided by the pooled estimate of the variance of that difference, is maximised by taking a as defined in (12.15). Consequently the discriminant rule given in (12.14) might be expected to perform reasonably well even when the assumption of normality is not wholly justified as, of course, it rarely is.

When more than two groups are involved it may be possible to determine *several* linear combinations of the original variables for separating groups. In general the number available is the smaller of p and $g - 1$; the new variables are referred to as *canonical* discriminant functions, the ith of which is the linear combination for which there is maximal separation between groups, in the sense stated above, subject to being uncorrelated with the $i - 1$ preceding functions. The coefficients defining these functions are found from the eigenvectors of the matrix $\mathbf{W}^{-1}\mathbf{B}$, where $\mathbf{W}$ is the within-group matrix of sums of squares and cross products and $\mathbf{B}$ the corresponding between-groups matrix. The eigenvalues of this matrix give the ratio of the between-groups sum of squares to within-group sum of squares for the corresponding discriminant function. Prior probabilities, when they are known, can be introduced into the discrimination rule as suggested in Section 12.2 (see equation (12.12)). In some cases these prior probabilities might be approximated well from knowledge of the relative sizes of the two populations. When little is known about the relative population sizes it is usual to use equal priors, i.e. to let $\Pi_1 = \Pi_2 = \frac{1}{2}$.

It is of some interest to note that it is possible to obtain the coefficients defining the linear discriminant function (i.e. $\mathbf{a}$ in 12.15), by using a regression approach. A dummy variable, y, is introduced which takes on a different value for members of D_1 than for members of D_2. If the two data sets are then treated as a single sample of size $n_1 + n_2$ the coefficients in the regression of y on $\mathbf{x}$ are proportional to $\mathbf{S}^{-1}(\bar{\mathbf{x}}_1 - \bar{\mathbf{x}}_2)$. (See T.A. Anderson, 1958.)

Any two values could be used to identify group membership but the most convenient are $n_2/(n_1 + n_2)$ for observations from D_1 and $-n_1/(n_1 + n_2)$ for observations from D_2. The average of the dummy variable over the whole data set is now zero, and a future individual is allocated to D_1 or D_2 according to whether the corresponding predicted value of y is positive or negative.

Under the normality assumption and with equal covariance matrices, the hypothesis that the p variables do not discriminate between the two populations can be stated as

$$H_0 : \boldsymbol{\mu}_1 = \boldsymbol{\mu}_2 \tag{12.16}$$

This can be tested (see Rao, 1965) by computing Hotelling's T^2 statistic (see Section 11.6).

To illustrate the use of Fisher's linear discriminant function we shall use the data shown in Table 12.1. These data arise from a large psychiatric study in which normal people and people diagnosed by a psychiatrist as ill are asked a series of questions relating to feelings of inadequacy, tension etc. The objective was to derive a classification rule which might be used to classify future subjects into one of the two classes. The sample mean vectors of the two groups and the pooled within-group covariance matrix are shown in Table 12.2.

Fisher's discriminant function is given by

$$-0.72x_1 - 0.37x_2 + 0.02x_3 - 0.34x_4 - 0.41x_5 \tag{12.17}$$

and the classification rule in (12.14) becomes, allocate an individual with scores x_1, x_2, x_3, x_4 and x_5 to the 'ill' group if

$$-0.72x_1 - 0.37x_2 + 0.02x_3 - 0.34x_4 - 0.41x_5 + 4.20 > 0 \tag{12.18}$$

The rule makes reasonable intuitive sense with those people scoring high on each variable likely to be designated 'well' and those scoring low 'ill'; variable 3 appears to contribute little to the discrimination. A plot of the discriminant scores for all individuals (see Figure 12.2) shows that there is only a little overlap between the two groups.

How might we evaluate the performance of the discriminant function more formally? One obvious way is to apply it to the original data and assess the misclassification rate, and Table 12.3 shows the number of subjects in the original samples correctly and incorrectly classified by (12.18). The misclassifi-

Table 12.1 Psychiatric data

Patient	x_1	x_2	x_3	x_4	x_5
1	2	2	2	2	2
2	2	2	2	1	2
3	1	1	2	1	1
4	2	2	2	1	2
5	1	1	2	1	2
6	1	1	2	1	1
7	2	2	2	2	2
8	1	1	2	1	2
9	1	1	2	1	2
10	2	1	2	1	2
11	2	2	2	1	2
12	2	1	2	1	2
13	1	1	2	2	2
14	1	1	2	1	2
15	3	3	2	3	2
16	4	3	3	3	2
17	3	3	2	3	3
18	3	2	2	3	2
19	4	2	2	2	2
20	2	3	2	3	3
21	2	2	2	2	3
22	3	2	2	1	3
23	3	3	2	1	3
24	2	2	2	2	2
25	3	1	3	4	4
26	2	2	3	1	2
27	3	2	2	4	2
28	3	2	2	3	3
29	2	2	2	3	1
30	3	2	4	3	3
31	3	1	3	1	3
32	1	2	2	1	2
33	3	3	2	4	3
34	2	3	2	4	3
35	3	3	3	4	3
36	2	1	2	3	3
37	4	4	4	4	4
38	2	1	2	3	3
39	4	1	4	4	4
40	3	3	2	2	3
41	2	2	2	1	2
42	4	2	2	2	2
43	3	3	2	3	3
44	2	3	2	2	2
45	4	3	1	2	3

(The first 15 patients form the 'ill' group, the remaining 30 the 'well' group).

x_1: Have you recently felt that you are playing a useful part in things?

x_2: Have you recently felt contented with your lot?

x_3: Have you recently felt capable of making decisions about things?

x_4: Have you recently felt that you're not able to make a start on anything?

x_5: Have you recently felt yourself dreading everything you have to do?

Each variable is coded 1 to 4; variables 1 to 3 are scored in the direction 'No' to 'Yes' and variables 4 and 5, 'Yes' to 'No'.

Table 12.2   Mean vectors and covariances matrix for psychiatric data

	Variable				
	x_1	x_2	x_3	x_4	x_5
Ill	1·60	1·47	2·00	1·40	1·93
Well	2·80	2·27	2·33	2·60	2·70
$S =$	0·57				
	0·19	0·55			
	0·12	0·04	0·34		
	0·23	0·15	0·16	0·90	
	0·18	0·07	0·14	0·25	0·40

Table 12.3   Performance on Fisher's discriminant function on psychiatric data as assessed on original sample

		Group allocated by discriminant rule	
		Ill	*Well*
Actual group	*Ill*	14	1
	Well	5	25

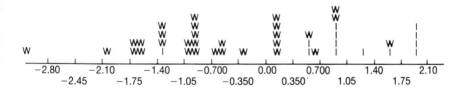

Figure 12.2   Discriminant scores for psychiatric patients

cation rate of 13% is relatively low, but estimated in this way (i.e. by applying the derived discriminant function to the original data), is likely to be a *highly optimistic* estimate of the actual misclassification rate.

A more realistic estimate of the misclassification rate may be obtained in a variety of ways (see Hand 1986 for details). Perhaps the most straightforward of these is the so-called 'leaving one out method,' where a discriminant function is derived on the basis of $(n-1)$ of the subjects and used to classify the individual not included, the whole process being repeated for each individual. Applying such an approach to the example discussed above, changes the values in Table 12.3 to those in Table 12.4. The estimated misclassification rate has now risen to 23%.

Apart from the assumption of normality the discriminant function in (12.14) assumes that the two populations of interest have equal covariance matrices.

Table 12.4 Performance on Fisher's discriminant function on psychiatric data as assessed by leaving one out method

		Group allocated by discriminant rule	
		III	*Well*
Actual group	*III*	14	1
	Well	9	21

When this is not the case the procedure outlined in Section 12.2 leads to a rule of the form, assign an individual with vector of scores **x** to D_1 if

$$\mathbf{x}'(\mathbf{S}_2^{-1} - \mathbf{S}_1^{-1})\mathbf{x} - 2\mathbf{x}'(\mathbf{S}_2^{-1}\bar{\mathbf{x}}_2 - \mathbf{S}_1^{-1}\bar{\mathbf{x}}_1)$$

$$+(\bar{\mathbf{x}}_2'\mathbf{S}_2^{-1}\bar{\mathbf{x}}_2 - \bar{\mathbf{x}}_1'\mathbf{S}_1^{-1}\bar{\mathbf{x}}_1) \geq \ln(|\mathbf{S}_2|/|\mathbf{S}_1|) + 2\ln(\Pi_1/\Pi_2) \qquad (12.19)$$

where $\mathbf{S}_1$ and $\mathbf{S}_2$ are the estimated covariance matrices of each group. Since the left-hand side of (12.19) contains square and cross-product terms, it is known in this case as a *quadratic discriminant function*. Its most important area of application arises when population means are considered equal making the linear discriminant function of no use.

When more than two populations are involved the methods described previously in this section can easily be extended to provide an appropriate classification rule. For example, with three groups the allocation rule would be based on three functions:

$$h_{12}(\mathbf{x}) = (\bar{\mathbf{x}}_1 - \bar{\mathbf{x}}_2)'\mathbf{S}^{-1}[\mathbf{x} - \frac{1}{2}(\bar{\mathbf{x}}_1 + \bar{\mathbf{x}}_2)] \qquad (12.20)$$

$$h_{13}(\mathbf{x}) = (\bar{\mathbf{x}}_1 - \bar{\mathbf{x}}_3)'\mathbf{S}^{-1}[\mathbf{x} - \frac{1}{2}(\bar{\mathbf{x}}_1 + \bar{\mathbf{x}}_3)] \qquad (12.21)$$

$$h_{23}(\mathbf{x}) = (\bar{\mathbf{x}}_2 - \bar{\mathbf{x}}_3)'\mathbf{S}^{-1}[\mathbf{x} - \frac{1}{2}(\bar{\mathbf{x}}_2 + \bar{\mathbf{x}}_3)] \qquad (12.22)$$

The classification rule now becomes, allocate an individual with vector of scores **x** to

$$D_1 \text{ if } h_{12}(\mathbf{x}) > 0 \text{ and } h_{13}(\mathbf{x}) > 0,$$
$$D_2 \text{ if } h_{12}(\mathbf{x}) < 0 \text{ and } h_{23}(\mathbf{x}) > 0,$$
$$D_3 \text{ if } h_{13}(\mathbf{x}) < 0 \text{ and } h_{23}(\mathbf{x}) < 0$$

Fisher's linear discriminant function is optimal when the two populations have multivariate normal distributions with equal covariance matrices, but has been shown to be relatively robust to departures from normality (see for example, Gilbert, 1968, Moore, 1973 and Krzanowski, 1977). In some cases, however, where the normality assumption is considered invalid it might be worth considering alternative methods; one which has been widely used is *logistic discrimination*.

12.3.2   Logistic Discrimination

The essential feature of this approach is to assume the following form for the probabilities of class membership (we shall assume we have two groups):

$$P(D_1|\mathbf{x}) = \exp(\alpha_o + \sum_{i=1}^{p} \alpha_i x_i)/[1 + \exp(\alpha_o + \sum_{i=1}^{p} \alpha_i x_i)] \qquad (12.23)$$

$$P(D_2|\mathbf{x}) = 1/[1 + \exp(\alpha_0 + \sum_{i=1}^{p} \alpha_i x_i)] \qquad (12.24)$$

The parameters, $\alpha_0, \alpha_1, \ldots, \alpha_p$ may be estimated by maximum likelihood using the procedure described in Section 9.4 (see Anderson 1972). The important point is that the estimation process is *independent* of the form assumed for the class density functions. Day and Kerridge (1967) and Anderson (1972) show that the method of discrimination has optimal properties under a wide range of assumptions about the underlying distributions including those relevant when both continuous *and* categorical variables are used to describe each individual.

After estimation of the parameters, allocation of new individuals can be performed on the basis of scores given by

$$\hat{\alpha}_0 + \sum_{i=1}^{p} \hat{\alpha}_i x_i \qquad (12.25)$$

If this is positive the individual is allocated to D_1 (since $P(D_1) > P(D_2)$), if negative to D_2 (assuming equal prior probabilities for the two groups).

As our first illustration of this technique we shall return to the pychiatric data discussed in Section 12.3.1. The linear discriminant function constructed for these data assumes that the 5 variables describing each patient have a multivariate normal distribution. A cursory examination of the data and the scoring scheme for the variables shows that this assumption is unlikely to be true. Consequently, it becomes of interest to compare the results of applying logistic discrimination to those obtained using the classical approach. Table 12.5 shows the maximum likelihood estimates of the parameters in the model specified by (12.23) and (12.24), and their standard errors. The classification rule given by this approach is, allocate an individual with scores x_1, x_2, x_3, x_4, x_5 to the ill group if

$$-0.93x_1 - 1.10x_2 - 2.65x_3 - 0.89x_4 - 0.96x_5 + 12.74 > 0 \qquad (12.26)$$

Comparing this to the discriminant rule in (12.18) shows that both are similar with subjects having low scores on the variables being allocated to the ill group. There are differences, however, particularly the greater importance of x_3 in the logistic discrimination. Here a total of five patients are incorrectly allocated.

As a further example of logistic regression we shall apply the technique to the data shown in Table 12.6 taken from Hand (1981b). These data were collected in a study designed to investigate whether the outcome of treatment of enuretic children with an alarm buzzer could be predicted from certain observations on the children. The two groups involved were

Table 12.5   Results of logistic discrimination applied to psychiatric data

Parameter	Estimate	Standard error
α_0	12·74	5·40
α_1	−0·93	0·83
α_2	−1·10	0·93
α_3	−2·65	1·99
α_4	−0·90	0·62
α_5	−0·96	0·95

(a) D_1=failure or relapse after apparent cure;
(b) D_2= long-term cure.

The estimates of the parameters and their standard errors are shown in Table 12.7.   From these the allocation rule becomes, allocate patient with scores $x_1, \ldots, x_5$ to D_1 if

$$0.74x_1 - 0.57x_2 + 0.05x_3 + 0.14x_4 + 0.01x_5 - 0.62 > 0 \qquad (12.27)$$

Again we may assess the performance of the rule (albeit optimistically) by comparing actual and predicted grouping on the original data.   These are shown in Table 12.8.   The misclassification rate is 33% indicating that the derived classification rule is only partly successful.

12.3.3   Other Methods

Although Fisher's discriminant function and logistic discrimination are the methods most commonly used to derive classification rules, a number of other techniques have been suggested for the same purpose. For example, several authors have suggested a 'nearest neighbour' rule which consists of finding the k nearest individuals (in terms of some metric), to the individual to be allocated. The new individual is then allocated to the group to which the majority of these neighbours belong. This rule may be easily modified to account for unequal prior probabilities.

Other examples of discrimination procedures are *classification trees* (see Breiman *et al.*, 1984), and *kernel methods* (see Hand, 1982). A comprehensive review of all such methods is given in Hand (1981b).

12.4   Selecting Variables

In many cases when using discriminant analysis, a question of interest is whether some subset of the original variables could provide a classification rule equal to or perhaps only marginally inferior to the rule based on all the variables. There are several different procedures available for selecting such subsets of variables. The general principle behind these methods is to choose a measure of separability between the groups and then either sequentially accumulate the variables which maximise the measure, or, beginning with all available variables, sequentially eliminate those whose removal leads to the least reduction in separation.   With classical discriminant analysis the

Table 12.6 Enuresis data (Taken with permission from Hand, 1981)

Patient	Group	x_1	x_2	x_3	x_4	x_5
1	D_2	0	0	0	1	0
2	D_2	1	0	0	1	0
3	D_2	0	0	0	1	0
4	D_2	0	0	0	0	0
5	D_2	0	0	0	1	0
6	D_1	0	0	0	1	0
7	D_1	1	0	0	1	0
8	D_1	0	0	0	1	1
9	D_1	1	0	0	1	1
10	D_1	1	0	0	1	0
11	D_1	1	0	0	1	0
12	D_2	0	0	0	1	0
13	D_1	1	0	0	0	0
14	D_1	1	0	1	1	0
15	D_2	1	0	0	1	0
16	D_2	1	0	0	1	0
17	D_1	0	0	1	1	0
18	D_2	0	0	1	1	0
19	D_2	0	0	1	1	0
20	D_1	1	0	0	0	0
21	D_1	0	0	0	1	0
22	D_2	0	0	1	1	0
23	D_1	1	0	0	1	0
24	D_2	1	0	0	1	0
25	D_1	1	0	0	1	1
26	D_2	0	0	0	0	0
27	D_2	1	0	1	0	0
28	D_1	1	0	0	1	0
29	D_1	1	1	0	1	1
30	D_1	0	0	1	1	0
31	D_2	0	0	1	1	0
32	D_1	1	0	1	1	0
33	D_2	0	0	1	1	0
34	D_1	1	0	1	1	0
35	D_2	0	0	0	1	1
36	D_1	0	0	1	1	0
37	D_1	1	0	0	1	0
38	D_2	1	0	1	1	0
39	D_2	1	0	1	1	0
40	D_2	1	0	1	1	0
41	D_2	0	1	1	1	0
42	D_1	1	0	1	1	0
43	D_1	0	0	0	1	0
44	D_1	0	0	0	1	0
45	D_2	1	0	0	1	1
46	D_1	0	0	0	1	0
47	D_1	1	0	1	1	1
48	D_1	1	0	1	0	1
49	D_2	0	0	0	1	0
50	D_2	0	0	0	1	0

Key

x_1 Whether or not there were family background difficulties (1 = yes, 0 = no).

x_2 Whether wetting occurred during the day (1 = yes, 0 = no).

x_3 The child's age (0 if greater than 8, 1 otherwise)

x_4 Whether the family had access to an inside w.c. (1 = yes, 0 = no).

x_5 Whether the child shared a room with more than one sibling (1 = yes, 0 = no).

Table 12.7   Results of logistic discrimination
applied to enuresis data in Table 12.6

Parameter	Estimate	Standard error
α_0	−0·52	0·52
α_1	0·74	0·21
α_2	−0·57	0·48
α_3	−0·05	0·21
α_4	0·14	0·34
α_5	−0·01	0·25

Table 12.8   Performance of logistic discrimination on enuresis data assessed on
original sample

Group allocated by discriminant rule			
		D_1	D_2
Actual group	D_1	34	21
	D_2	16	42

criterion generally used to assess variables for entry (or removal) is their contribution to Hotelling's T^2, as judged by the appropriate F-statistic. In logistic regression the criterion used is one based on a ratio of likelihoods of the current and candidate models. Apart from details, however, the methods used here are essentially equivalent to the stepwise procedures previously described in connection with regression in Chapter 8.

As an example of the use of these procedures in the context of discriminant analysis we shall again use the enuresis data in Table 12.5. Applying first logistic regression and beginning with all five variables in the model, variables are removed sequentially as shown in Table 12.9. The increases in chi-square values as variables are removed are very small, and the only variable that appears to be useful in the discrimination between children for whom the buzzer treatment fails and those for whom it works is family background; those children exposed to family background problems are those for whom the treatment is likely to fail.

A similar conclusion is reached by applying a stepwise procedure to Fisher's discriminant function derived from these data.

12.5   Summary

Discriminant analysis techniques produce rules which can be used to classify new individuals whose group membership is unknown. The two most commonly used methods are Fisher's linear discriminant function, and logistic discrimination. The former is optimal when the classes have multivariate densities with the same covariance matrix, and the latter is applicable to a wide range of density functions. Excellent detailed accounts of the area are given by Hand (1981b) and Gnanadesikan and Kettenring (1989).

Table 12.9 Selecting variables in logistic discrimination as applied to enuresis data

Step number	Variables included	Goodness of fit chi-square
0	x_1, x_2, x_3, x_4, x_5	9·923
1	x_1, x_2, x_3, x_4	9·925
2	x_1, x_2, x_4	9·978
3	x_1, x_2	10·158
4	x_1	12·046

x_1 cannot be removed because it would cause a
significant increase in the chi-square goodness-of-
fit measure. (x_1–x_5 are identified in Table 12.6)

Table 12.10

Skull	Early predynastic				Late predynastic			
	x_1	x_2	x_3	x_4	x_1	x_2	x_3	x_4
1	131	138	89	49	124	138	101	48
2	125	131	92	48	133	134	97	48
3	131	132	99	50	138	134	98	45
4	119	132	96	44	148	129	104	51
5	136	143	100	54	126	124	95	45
6	138	137	89	56	135	136	98	52
7	130	130	108	48	132	116	100	54
8	125	136	93	48	133	130	102	48
9	131	134	102	51	131	134	96	50
10	134	134	99	51	133	125	94	46

x_1 – maximum breadth, x_2 – basibregmatic height, x_3 = basialvcolar
length, x_4 = nasal height (all in mm). Reprinted with permission from
Chapman and Hall from Manley 1986

Exercises

12.1 Table 12.10 (taken with permission form Manly, 1986) shows measurements made on 10 male Egyptian skulls form two epochs. Construct
a discriminant function based on the assumption that the data from
each population have a multivariate normal distribution with the same
covariance matrix. Into which of the two groups would a skull with
measurements $x_1 = 127$, $x_2 = 129$, $x_3 = 95$, $x_4 = 51$ be placed by the
derived discriminant function?

12.2 In the two group discrimination problem suppose that

$$f_i = \binom{n}{x} p_i^x (1 - p_i)^{n-x} \qquad 0 < p_i < 1, \quad i = 1, 2$$

where p_1 and p_2 are known. Π_1 and Π_2 are the usual prior probabilities
derive the classification rule based on equation (12.1).

12.3 For the data in Exercise 12.1 is it possible to discriminate between the
two sets of skulls on the basis of measurements of maximum breadth and
basibregmatic height alone?

Table 12.11

Voting intention	Age	Sex	Social class
C	46	M	UMC
C	23	M	WC
C	72	F	UMC
C	52	F	LMC
C	33	F	UMC
C	19	M	WC
L	22	M	LMC
L	49	M	LMC
C	67	F	LMC
C	50	F	LMC
C	42	F	LMC
L	21	F	UMC
L	28	M	WC
L	26	M	WC
L	75	M	WC
L	41	F	LMC
L	35	F	WC
C	41	M	UMC
C	48	M	UMC
L	32	M	WC
L	35	M	WC
C	55	F	LMC
C	47	F	UMC
C	19	F	UMC
L	24	M	WC

C = conservative, L = labour, UMC = upper middle class, LMC = lower middle class, WC = working class.

12.4 In a survey respondents were asked for which political party, Labour or Conservative, they intended to vote in the next General Election. Each respondent was also asked their age, and their sex and social class were noted by the interviewer. The data collected for 25 individuals is given in Table 12.11. Construct a logistic discriminant function for predicting voting intention from age, sex and social class (which may have to be recoded).

PART IV
LATENT VARIABLE MODELS

One of the key assumptions in fitting linear regression models such as those described in Chapter 8 is that all of the variables other than the response variable have been measured without error. Clearly, in areas such as sociology and psychology this is very rarely the case. More importantly, perhaps, research workers in the social sciences are often also interested in modelling variation and covariation of traits that by definition cannot be measured directly. These so-called, *latent variables* are often theoretical concepts such as intelligence, psychoticism or social class, so that, in practice, we have to make measurements on manifest variables that are assumed to be indicators for the measures that one is really interested in.

In the last two chapters of this text we introduce statistical methods that are suitable for the exploration of the relationships between latent variables and manifest variables or between sets of latent variables, where the latent variables can represent concepts that cannot be measured directly or measurements that cannot be made without error.

Factor analysis is essentially a regression model for the observed variables on the unobserved latent variables or factors. *Structural equation models* are more general and are specified in terms of tentative causal relations between a set of latent dependent and latent independent variables. In both cases one is interested in exploring or describing the structure of a covariance matrix.

13

Factor Analysis

13.1 Introduction

In many areas of psychology, sociology and the like it is sometimes not possible to measure directly the concepts that are of major interest. Two obvious examples are intelligence and social class. In such cases the researcher will often collect information on variables likely to be indicators of the concepts in question and then try to discover whether the relationships between these observed variables are consistent with them being measures of a single underlying latent variable or whether some more complex structure has to be postulated. For example, the psychologist interested in intelligence may record the examination scores for a number of students in a variety of different subjects and the sociologist attempting to assess social class may note a person's occupation, educational background, whether or not they own their home, and so on. In such studies the most frequently used method of analysis is some form of *factor analysis*, a term which subsumes a fairly large variety of procedures, all of which have the aim of ascertaining whether the interrelations between a set of observed variables are explicable in terms of a small number of underlying, unobservable variables or *factors*. The basic factor analysis model is essentially the same as that of multiple regression discussed in Chapters 7 and 8 except that, here, the observed variables are regressed on the unobservable factors.

13.2 The Basic Factor Analysis Model

Factor analysis is concerned with whether the covariances or correlations between a set of observed variables, $\mathbf{x}' = [x_1,\ldots,x_p]$ can be 'explained' in terms of a smaller number of unobservable, latent variables $f_1,\ldots,f_k$, where $k < p$. Explanation in this case means that the correlation between each pair of observed variables results from their mutual association with the latent variables; consequently the *partial* correlations between any pair of observed variables given the values of $f_1,\ldots,f_k$ should be approximately zero. The smallest value of k compatible with this aim gives the most *parsimonious* explanation. The simplest model that satisfies the requirement that the observed variables are conditionally uncorrelated, given the values of all the f_i, is the following:

$$x_1 = \lambda_{11}f_1 + \lambda_{12}f_2 + \cdots + \lambda_{1k}f_k + u_1$$
$$x_2 = \lambda_{21}f_1 + \lambda_{22}f_2 + \cdots + \lambda_{2k}f_k + u_2 \qquad (13.1)$$
$$\vdots$$
$$x_p = \lambda_{p1}f_1 + \lambda_{p2}f_2 + \cdots + \lambda_{pk}f_k + u_p$$

or written more concisely

$$\mathbf{x} = \mathbf{\Lambda f} + \mathbf{u} \qquad (13.2)$$

where

$$\mathbf{\Lambda} = \begin{pmatrix} \lambda_{11} & \cdots & \lambda_{1k} \\ \vdots & & \vdots \\ \lambda_{p1} & \cdots & \lambda_{pk} \end{pmatrix}$$

and

$$\mathbf{f} = \begin{pmatrix} f_1 \\ \vdots \\ f_k \end{pmatrix}$$

and

$$\mathbf{u} = \begin{pmatrix} u_1 \\ \vdots \\ u_p \end{pmatrix}$$

(We have assumed that the mean vector of $\mathbf{x}$ is the null vector; this is of no practical consequence since we are essentially interested only in the covariance or correlational structure of the variables.) To ensure that the x_i are independent given the f_i we need the u_i to be uncorrelated with each other and with the f_i. Since the latter are unobserved and therefore give no information on location or scale we shall assume they occur in standardised form and for the present we shall also assume that they are independent of one another; the last assumption will be relaxed in later discussions. With these assumptions the factor model given by equation (13.2) implies that the variance σ_i^2 of the variable x_i is given by

$$\sigma_i^2 = \sum_{j=1}^{k} \lambda_{ij}^2 + \psi_i \qquad (13.3)$$

where ψ_i is the variance of u_i; the covariance σ_{ij} of variables x_i and x_j is given by

$$\sigma_{ij} = \sum_{l=1}^{k} \lambda_{il}\lambda_{jl} \qquad (13.4)$$

Equation (13.3) demonstrates that the factor model implies that the variance of an observed variable can be split into two parts. The first

$$h_i^2 = \sum_{j=1}^{k} \lambda_{ij}^2 \qquad (13.5)$$

is known as the *communality* of the variable and represents the variance shared with the other variables via the common factors. The second part, ψ_i, is called the *specific* or *unique* variance and relates to the variability in x_i not shared with other variables. Equation (13.4) shows that the covariance of two observed variables depends solely on their relationship with the common factors as indicated by the appropriate factor loadings, which, in the case of orthogonal factors, give the correlations between factors and observed variables. In (13.4) the specific factors play no part in determining σ_{ij}. Equations (13.3) and (13.4) can be summarised in the form

$$\Sigma = \Lambda\Lambda' + \Psi \tag{13.6}$$

where

$$\Psi = \text{diag}(\psi_i)$$

and Σ is the population covariance matrix of x. So the k-factor model given by equation (13.2) implies that the covariance matrix of the observed variables has the form indicated in (13.6). The converse also holds; if Σ can be decomposed into the form in (13.6) then the k-factor model holds for x. In practice, Σ will be estimated by the sample covariance matrix, S, and from this we shall want to obtain estimates of Λ and Ψ which might then be used to ascertain whether a k-factor model is appropriate for the data under investigation (see Section 13.4). Unfortunately the factor loading matrix Λ is not uniquely determined by equations (13.2) and (13.6). This is easily seen if we suppose M is any orthogonal matrix of order $(k \times k)$, and we rewrite (13.2) as

$$x = (\Lambda M)(M'f) + u \tag{13.7}$$

This satisfies all the requirements of a k-factor model as outlined above, with new factors, $f^* = M'f$, and new factor loadings ΛM. Consequently (13.7) implies that

$$\Sigma = (\Lambda M)(\Lambda M)' + \Psi \tag{13.8}$$

which reduces to (13.6) since $MM' = I$. This implies that if the factors f with loadings Λ provide an explanation for the observed covariances of the x_i, then so also do the factors f^*, with the loadings ΛM, for any orthogonal matrix M. It becomes necessary, therefore, to impose a set of constraints on the parameters of the model, to ensure a unique solution. In general this is achieved by requiring the matrix G given by

$$G = \Lambda'\Psi^{-1}\Lambda \tag{13.9}$$

to be diagonal, with its elements arranged in descending order of magnitude. This constraint leads to factors such that the first one makes the maximum contribution to the common variance in the elements of x, the second makes the maximum contribution subject to being uncorrelated with the first and so on. (Compare principal components analysis in Chapter 4.) Although the constraints introduced by equation (13.9) are necessary to achieve a unique solution for the factor model, they are, essentially, arbitrary, and it may be that having obtained parameter estimates using them, a more interpretable solution can be achieved using the transformed model with loadings $\Lambda^* = \Lambda M$. Such a process, which is generally known as *factor rotation*, is discussed in detail in Section 13.5.

13.3 Estimating the Parameters in the Factor Model

The estimation problem in factor analysis is essentially that of finding estimates $\hat{\Lambda}$ and $\hat{\Psi}$ satisfying the constraints (13.9), and for which

$$S \approx \hat{\Lambda}\hat{\Lambda}' + \hat{\Psi} \qquad (13.10)$$

The type of solution possible depends on the difference between the number of independent elements of S and the number of free parameters in the factor analysis model; the former is given by $\frac{1}{2}p(p+1)$ and the latter by $p + pk - \frac{1}{2}k(k-1)$, which arises from counting the p residual variances and the pk factor loadings and subtracting the $\frac{1}{2}k(k-1)$ constraints imposed by (13.9). The difference s is therefore given by

$$s = \frac{1}{2}p(p+1) - [p + pk - \frac{1}{2}k(k-1)] \qquad (13.11)$$

$$= \frac{1}{2}[(p-k)^2 - (p+k)] \qquad (13.12)$$

Three cases need to be considered:

(1) $s < 0$. Here there are fewer equations than free parameters, and so an infinite number of exact solutions are possible. Clearly the factor model is not well defined in this situation.

(2) $s = 0$. The factor model contains as many parameters as elements of S, and the model offers no simplification of the relationships amongst the observed variables. A unique solution may be found but not necessarily one with all the specific variances greater than zero.

(3) $s > 0$. This is the usual situation and the only one of real interest. Here there will be fewer parameters in the factor model than there are elements of S; consequently the model may provide a simpler explanation of the relationships amongst the observed variables than is provided by the elements of S.

   In the last case an exact solution giving $S = \hat{\Lambda}\hat{\Lambda}' + \hat{\Psi}$ is not possible and methods have been developed for giving approximate solutions. Before discussing these methods, however, we shall consider a simple example, originally discussed by Spearman (1904), where $s = 0$ and an exact solution is possible. Spearman considered children's examination marks in three subjects, Classics (x_1), French (x_2) and English (x_3) and found the following correlations:

		Classics	French	English
	Classics	1.00		
$R =$	French	0.83	1.00	
	English	0.78	0.67	1.00

If we assume that a single underlying factor, f, is adequate for explaining these correlations then the appropriate factor model is of the form

$$x_1 = \lambda_1 f + u_1,$$
$$x_2 = \lambda_2 f + u_2, \qquad (13.13)$$
$$x_3 = \lambda_3 f + u_3.$$

In this example the common factor, f, might be equated with intelligence or general intellectual ability, and the specific factors, u_i, will have small variances if the corresponding observed variable is closely related to f. The value of s in (13.12) is zero and by equating elements of the observed correlation matrix to those implied by the single factor model we are able to find estimates of $\lambda_1, \lambda_2, \lambda_3, \psi_1, \psi_2,$ and ψ_3 such that the model fits exactly. The resulting equations are

$$\lambda_1\lambda_2 = 0.83, \lambda_1\lambda_3 = 0.78, \lambda_2\lambda_3 = 0.67,$$
$$\psi_1 = 1.0 - \lambda_1^2, \psi_2 = 1.0 - \lambda_2^2, \psi_3 = 1.0 - \lambda_3^2$$

leading to

$$\lambda_1 = 0.99, \quad \lambda_2 = 0.84, \quad \lambda_3 = 0.79$$
$$\psi_1 = 0.02, \quad \psi_2 = 0.30, \quad \psi_3 = 0.38$$

Suppose now that the observed correlations had been

	Classics	French	English
Classics	1.00		
$\mathbf{R} =$ French	0.84	1.00	
English	0.60	0.35	1.00

leading to the solution

$$\lambda_1 = 1.2, \quad \lambda_2 = 0.7, \quad \lambda_3 = 0.5$$
$$\psi_1 = -0.44 \quad \psi_2 = 0.51 \quad \psi_3 = 0.75$$

Clearly the solution is unacceptable because of the negative variance estimate. In the examples above we have first found estimates of the factor loadings and then used

$$\hat{\psi}_i = s_i^2 - \sum_{j=1}^{k} \hat{\lambda}_{ij}^2 \tag{13.14}$$

to find estimates of specific variances, where s_i^2 is the sample variance of variables x_i (or unity if a correlation matrix is being used). The estimation procedures to be described in the next two sections, namely *principal factor analysis* and *maximum likelihood factor analysis*, employ the same principle.

13.4 Principal Factor Analysis

Principal factor analysis is essentially equivalent to a principal components analysis (see Chapter 4) performed on the *reduced covariance matrix*, $\mathbf{S}^*$, obtained by replacing the observed diagonal elements of $\mathbf{S}$ with estimated communalities. Two frequently used estimates of the latter of:

(1) the square of the multiple correlation coefficient of the ith variable with all other variables (see Chapter 7);
(2) the largest of the absolute values of the correlation coefficients between the ith variable and one of the other variables.

Each of these estimates will give higher communality values when x_i is highly correlated with the other variables, which is what is required. Next a principal components analysis is performed on $\mathbf{S}^*$ and the first k components used to provide estimates of the loadings in the k-factor model. Revised estimates of the specific variances are found from (13.14) and a principal factor analysis is considered satisfactory if all of these are non-negative. Although many users might be satisfied with the parameter estimates obtained by the procedure outlined above, it is possible to compute revised communality estimates from $\hat{\Lambda}$, use these to find new estimates of the specific variances and continue this iterative process until the differences between successive communality estimates are negligible. Difficulties can arise, however, if at any time communality estimates exceed the variance of the corresponding observed variable, since this would give rise to negative estimates of the specific variances (see equation (13.14)). Some numerical examples of principal factor analysis appear in Section 13.3.3.

13.5   Maximum Likelihood Factor Analysis

If we assume that our raw data arise from a multivariate normal distribution, then we can apply the method of maximum likelihood to derive estimates of factor loadings and specific variances. The method was first applied to factor analysis by Lawley (1940, 1941, 1943) but its routine use had to await the development of computers and suitable numerical optimisation procedures. The latter were provided in the late 1960s, notably by Joreskog (1967), Joreskog and Lawley (1968) and Lawley (1967). The likelihood function that has to be maximised is

$$L = -\frac{1}{2}n[\ln|\Sigma| + \text{trace}[\mathbf{S}\Sigma^{-1}]] \qquad (13.15)$$

where $\mathbf{S}$ is the observed covariance or correlation matrix, and Σ is the corresponding matrix predicted by the k-factor model; since Σ is a function of Λ and Ψ so to is L. For various reasons it is more convenient to *minimise* the function F given by

$$F = \ln|\Sigma| + \text{trace}[\mathbf{S}\Sigma^{-1}] - \ln|\mathbf{S}| - p \qquad (13.16)$$

This is equivalent to maximising L since $L = -\frac{1}{2}nF + $ a function of the observations. The function F takes the value zero if $\Sigma = \mathbf{S}$ and values greater than zero otherwise. Details of how F is minimised are given in Lawley and Maxwell (1971), Mardia *et al.* (1979), and Everitt (1984, 1987). One of the main advantages of using the maximum likelihood method of estimation is that it enables us to test the hypothesis that k common factors are sufficient to describe the observed relationships in the data. This test will be described in Section 13.4. Here we move on to consider some numerical examples of both principal factor analysis (PFA) and maximum likelihood factor analysis (MLFA).

13.6   Numerical Examples of PFA and MLFA

To illustrate the application of the two estimation procedures described in the previous section we shall again use the data on drug usage rates described in Chapter 4. The results of fitting a two-factor model (procedures for choosing the number of factors will be described in Section 13.4) by both PFA and

Table 13.1 Two-factor solutions given by principal components analysis, principal factor analysis and maximum likelihood factor analysis on drug usage correlation matrix

Variable	PC 1	PC 2	PFA 1	PFA 2	MLFA 1	MLFA 2
1 cigs	0·58	0·40	0·54	0·30	0·57	0·20
2 beer	0·60	0·57	0·59	0·51	0·66	0·43
3 wine	0·55	0·56	0·54	0·48	0·61	0·42
4 liquor	0·66	0·46	0·65	0·40	0·71	0·31
5 cocaine	0·44	−0·41	0·38	−0·32	0·32	−0·35
6 tranquil	0·61	−0·37	0·57	−0·35	0·51	−0·43
7 drugstore	0·37	−0·27	0·31	−0·19	0·28	−0·22
8 heroin	0·42	−0·45	0·37	−0·35	0·31	−0·37
9 marijuana	0·71	0·23	0·67	0·16	0·68	0·03
10 hashish	0·69	−0·07	0·64	−0·10	0·62	−0·21
11 inhal	0·58	−0·24	0·52	−0·21	0·47	−0·27
12 hallu	0·52	−0·47	0·47	−0·41	0·41	−0·48
13 amphet	0·60	−0·33	0·65	−0·34	0·60	−0·45

MLFA are given in Table 13.1. For comparison this table also contains the first two principal components (see Chapter 4). The solutions in terms of factor loadings are clearly very similar. (The initial communality estimates for both PFA and MLFA were the squared multiple correlations of each variable with all other variables.) One simple way to assess how well the model fits the data is to examine the differences between the observed correlations and those predicted under the model; such *residual* correlation matrices for PFA and MLFA are shown in Table 13.2. The residual correlations from a principal components analysis are shown in Table 13.3. In each case the residual correlations are relatively small, although they are noticeably larger for the principal components analysis, a reflection of the fact that the factor analyses procedures are better equipped to account for the correlations between a set of observed variables rather than to simply explain their variances.

13.7 Determining the Number of Factors

An important question which arises when using factor analysis is 'How well does the model with a particular number of common factors fit the data?' This question has been considered by a number of workers who have suggested a variety of techniques for providing an answer. Many are fairly informal procedures based on experience and intuition rather than any formal sampling model. For example, one of the most popular criteria for indicating number of factors is to retain only the factors associated with eigenvalues greater than unity. This simple criterion appears to work well when applied to samples from artificially created population models, and generally gives results consistent with investigators' expectations (although this may not be a particularly objective judgement!).Another informal method for determining the number of factors is the so-called scree-test suggested by Cattell (1965). This method uses the graph of eigenvalues and chooses the number of factors corresponding to the point where the eigenvalues begin to level off to form an almost

Table 13.2

(a) Residual correlations from two-factor solution given by principal factor analysis

	1 cigs	2 beer	3 wine	4 liq	5 coc	6 tran
1 cigs	—					
2 beer	−0·027	—				
3 wine	−0·012	0·051	—			
4 liquor	−0·037	0·008	0·036	—		
5 cocaine	0·004	0·007	0·004	−0·003	—	
6 tranq	−0·001	−0·013	0·001	0·028	0·024	—
7 drugst	−0·020	0·017	0·036	−0·004	0·029	−0·023
8 heroin	−0·013	0·024	0·037	−0·002	0·070	0·026
9 marij	0·102	−0·037	−0·075	−0·022	−0·017	−0·009
10 hashish	−0·011	−0·009	−0·055	−0·008	0·030	−0·018
11 inhal	0·028	0·004	0·006	0·002	0·010	−0·041
12 hallu	−0·032	0·020	0·020	−0·002	−0·032	−0·043
13 amphet	−0·007	−0·013	−0·002	0·004	−0·079	0·056

	7 drugs	8 heroin	9 marij	10 hash	11 inhal	12 hallu
7 drugs	—					
8 heroin	0·019	—				
9 marj	−0·029	−0·037	—			
10 hash	−0·055	−0·050	0·122	—		
11 inhal	0·108	0·025	−0·012	−0·048	—	
19 hallu	0·005	0·002	−0·047	0·026	0·010	—
13 amphet	−0·038	−0·046	0·010	0·017	−0·017	0·061

(b) Residual correlations from two-factor solution given by maximum likelihood factor analysis

	1 cigs.	2 beer	3 wine	4 liq	5 coc	6 tran
1 cigs	—					
2 beer	−0·020	—				
3 wine	−0·012	0·028	—			
4 liquor	−0·029	−0·001	0·020	—		
5 cocaine	0·001	0·006	0·006	−0·005	—	
6 tranq	−0·004	−0·010	0·005	0·025	0·033	—
7 drugs	−0·024	0·012	0·032	−0·008	0·044	−0·014
8 heroin	−0·020	0·019	0·034	−0·008	0·090	0·035
9 marij	0·119	−0·022	0·064	−0·009	−0·019	−0·017
10 hashish	−0·006	−0·003	−0·050	−0·005	0·032	−0·028
11 inhal	0·028	0·003	0·005	0·001	0·026	−0·035
12 hallu	−0·039	0·021	0·023	−0·007	−0·020	−0·049
13 amphet	−0·008	−0·006	0·006	0·004	−0·075	0·040

	7 drugs	8 heroin	9 marij	10 hash	11 inhal	12 hallu
7 drugs	—					
8 heroin	0·034	—				
9 marij	−0·030	−0·043	—			
10 hash	−0·053	−0·050	0·124	—		
11 inhal	0·121	0·041	−0·011	−0·045	—	
12 hallu	0·015	0·014	−0·058	0·016	0·018	—
13 amphet	−0·033	−0·043	0·001	0·001	−0·015	0·047

horizontal straight line. Such a plot was described previously in Chapter 4 for use in association with principal components analysis. A more formal

Table 13.3 Residual correlations from two-component solution given by principal component analysis

	1 cigs	2 beer	3 wine	4 liq	5 coc	6 tran
1 cigs	—					
2 beer	−0·129	—				
3 wine	−0·126	−0·031	—			
4 liquor	−0·138	−0·053	−0·046	—		
5 cocaine	0·025	0·041	0·042	0·017	—	
6 tranq	−0·006	−0·011	0·006	0·022	−0·071	—
7 drugst	−0·015	0·036	0·057	0·003	−0·063	−0·105
8 heroin	−0·017	0·066	0·085	0·026	−0·049	−0·071
9 marij	0·006	−0·113	−0·160	−0·099	−0·027	−0·033
10 hash	−0·068	−0·053	−0·101	−0·056	0·027	−0·072
11 inhal	0·005	−0·006	−0·002	−0·017	−0·079	−0·121
12 hallu	−0·013	0·044	0·049	0·012	−0·141	−0·126
13 amphet	−0·023	−0·024	−0·011	−0·010	−0·158	0·001

	7 drugs	8 heroin	9 marij	10 hash	11 inhal	12 hallu
7 drugs	—					
8 heroin	−0·077	—				
9 marij	−0·048	−0·040	—			
10 hash	−0·110	−0·104	0·062	—		
11 inhal	0·032	−0·065	−0·053	−0·113	—	
19 hallu	−0·087	−0·112	−0·055	−0·024	−0·074	—
13 amphet	−0·111	−0·126	0·017	−0·03001	−0·086	−0·003

procedure is the large-sample chi-squared test associated with the maximum likelihood solution. From a statistical point of view this is the most satisfactory method, provided that the assumptions are adequately met. The form of the test statistic is as follows

$$U = n' \min(F) \qquad (13.17)$$

where $n' = n - 1 - \frac{1}{6}(2p + 5) - \frac{2}{3}k$ and F is given by equation (13.15). If k common factors are sufficient to describe the data then U is asymptotically distributed as a χ^2 with v degrees of freedom where

$$v = \frac{1}{2}(p - k)^2 - \frac{1}{2}(p + k) \qquad (13.18)$$

In most exploratory studies k cannot be specified in advance and so a sequential procedure is used to determine k. Starting with some small value for k (usually 1), the parameters in the factor model are estimated using the maximum likelihood method. If the test statistic, U, is not significant we accept the model with this number of factors, otherwise we increase k by one and repeat the process, until an acceptable solution is reached. If at any stage the degrees of freedom, v, becomes zero then either no non-trivial solution is appropriate, or alternatively the factor model itself with its assumption of linear relationships between observed and latent variables is questionable. As an example of this test, U was calculated for several factor models applied to the drug usage data. The values obtained were

$$\text{one-factor model} \quad U = 2287.26 \quad df = 65$$
$$\text{two-factor model} \quad U = 477.52 \quad df = 53$$

three-factor model $U = 230.21$ df $= 42$

four-factor model $U = 115.87$ df $= 32$

five-factor model $U = 66.44$ df $= 23$

All of these values are significant beyond the 5% level, indicating that more than five factors are needed to provide an adequate fit according to this particular criterion. In practice it is unlikely that any more complex model would be considered.

13.8   Rotation of Factors

The constraints on the factor loadings given by equation (13.9) were introduced to make the parameter estimates in the factor analysis model unique; they lead to orthogonal factors that are arranged in descending order of importance. These properties are not, however, inherent in the factor model, and merely considering only such solutions can lead to difficulties in interpreting factors. For example, two consequences of making these arbitrary impositions are: (1) that the factorial complexity of variables is likely to be greater than one, regardless of the underlying true model, that is variables will have substantial loadings on more than one factor; and (2) that, except for the first factor, the remaining factors may be bipolar — some variables have positive loadings on a factor while others have negative loadings. Both of these can make interpretation more difficult since this is most straightforward if each variable is highly loaded on at most one factor, and if all factor loadings are either large and positive or near zero, with few intermediate values. The variables are then split into disjoint sets, each of which is associated with one factor. Consequently after an initial factoring has determined the number of common factors necessary, and the communalities of each variable, a further step is taken in which the factor loadings are transformed by post multiplying the loading matrix by an orthogonal matrix. The new loading matrix will give an equal explanation of the relationships between the observed variables as described in Section 13.2. The advantage of the new loading matrix is that it may, if the transformation is well chosen, lead to a more easily interpretable solution. Such a transformation is equivalent to a rigid *rotation* of the axes of the originally identified factor space (see Chapter 2 for a simple description of the rotation of axes), and seeks to achieve a pattern of loadings which Thurstone (1931) referred to as *simple structure*; this has the following properties:

(1)   each row of Λ should contain at least one zero;

(2)   each column of Λ should contain at least k zeros;

(3)   every pair of columns of Λ should contain several responses whose loadings vanish in one column but not in the other;

(4)   if the number of factors is four or more, every pair of columns should contain a large number of responses with zero loadings in both columns;

(5)   conversely for every pair of columns of Λ only a small number of responses should have non-zero loadings in both columns.

The achievement of simple structure would mean essentially that the observed variables fall into mutually exclusive groups whose loadings are high

Table 13.4 Correlation coefficients between six school subjects

Subject	1	2	3	4	5	6
French	1·00					
English	0·44	1·00				
History	0·41	0·35	1·00			
Arithmetic	0·29	0·35	0·16	1·00		
Algebra	0·33	0·32	0·19	0·59	1·00	
Geometry	0·25	0·33	0·18	0·47	0·46	1·00

Table 13.5 Two-factor solution for correlation matrix in Table 13.4

Variable	Unrotated loadings 1	2	Rotated loadings 1	2
1 French	0·55	0·43	0·20	0·62
2 English	0·57	0·29	0·30	0·52
3 History	0·39	0·45	0·05	0·55
4 Arithmetic	0·74	−0·27	0·75	0·15
5 Algebra	0·72	−0·21	0·65	0·18
6 Geometry	0·59	−0·13	0·50	0·20

on single factors, perhaps moderate to low on a few factors and of negligible size on the remaining factors. For a two-factor model the process of rotation is straightforward since the loadings may be plotted and a pair of axes chosen which make the interpretation of the factors simpler. An example serves as the best description of this procedure and for this purpose we shall use the two-factor solution derived from the correlation matrix in Table 13.4. The initial factor loadings are shown in Table 13.5, and plotted in Figure 13.1. By referring each variable to the new axes shown, which correspond to a rotation of the original axes through 40 degrees, the new set of loadings, also shown in Table 13.5, are found. The two factors may now be reasonably labelled as *mathematical* and *verbal*.

Such graphical rotation is simple and efficient when dealing with two-factor solutions in which there are relatively clear clusters of variables. But whenever the pattern is not very clear or there are many factors to examine the method is no longer practicable. Consequently more formal methods for rotation have been developed, the most common of which is perhaps the varimax technique originally proposed by Kaiser (1958). This has as its rationale the provision of factors with a few large loadings and as many near-zero loadings as possible. This is accomplished by an iterative maximisation of a quadratic function of the loadings, details of which are given in Mardia *et al.* (1979). As an example, we shall consider the two-factor solution given by MLFA on the drug usage data. The varimax rotated factors are shown in Table 13.6. The labelling of these two factors as *hard drugs* and *soft drugs* is relatively clear, although marijuana and hashish have moderate loadings on both factors.

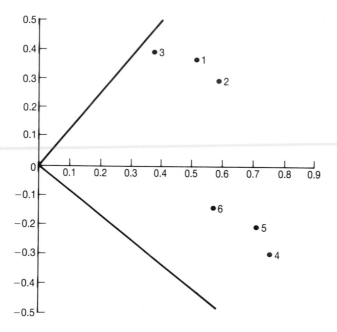

Figure 13.1   Rotation of factors

Table 13.6   Varimax solution for drug usage data

Drug	F_1	F_2
1 cigs	0·180	0·577
2 beer	0·052	0·794
3 wine	0·029	0·743
4 liquor	0·180	0·750
5 cocaine	0·474	0·047
6 tranq	0·650	0·154
7 drugst	0·339	0·092
8 heroin	0·486	0·024
9 marij	0·378	0·563
10 hash	0·535	0·369
11 inhal	0·498	0·220
12 hallu	0·628	0·044
13 amphet	0·725	0·211

In some cases factors become easier to interpret if we allow the axes representing them to become *oblique* rather than to remain orthogonal as with varimax rotation. Various methods for the oblique rotation of factors are available such as *oblimin* (see Nie *et al.* 1975) and these can be useful in particular circumstances. Because oblique solutions involve the introduction of

Table 13.7

	var 1	var 2	var 3	var 4	var 5	var 6	var 7	var 8
Factor 1	X	X	X	X	X	X	X	X
Factor 2	X	X	0	0	0	0	0	X
Factor 3	X	X	X	X	X	X	0	0

correlations between factors, a different type of complexity is introduced into the factor analysis solution. Factor rotation is often regarded as controversial since it apparently allows the investigator to impose on the data whatever type of solution is required. But this is clearly *not* the case since although the axes may be rotated about their origin, or may be allowed to become oblique, *the distribution of the points will remain invariant.* Rotation is simply a procedure which allows new axes to be chosen so that the positions of the points can be described as simply as possible. (It should be noted that rotation techniques are also often applied to the results from a principal components analysis in the hope that it will aid in their interpretability. Although in some cases this may be acceptable it does have several disadvantages which are listed by Jolliffe, 1989; the main one is that the defining property of principal components, namely that of accounting for maximal proportions of the total variation in the observed variables, is lost after rotation.)

13.9   Confirmatory Factor Analysis

The factor methods discussed in the previous sections may be described as exploratory, since they are concerned not only with the problem of determining the number of factors required for a data set, but also with the problem of rotation to facilitate the interpretation of factors. In some situations, however, an investigator, perhaps on the basis of previous research, may wish to postulate in advance the number of factors and the pattern of zero and non-zero loadings on them. Testing such a specific hypothesis about the factorial composition of a set of variables in some population involves specifying a number of parameters in the model as fixed and a number as free. Those designated are then estimated (generally by maximum likelihood methods), and the fit of the model assessed. For example, an investigator might postulate a three- factor model for a set of eight variables, where the factors have the pattern shown in Table 13.7.

Here the *X* represents free parameters to be estimated; the remaining loadings are considered fixed with, in this case, values of zero. If the factors were postulated to be orthogonal there would be a further six fixed parameters in the model, since the elements of the correlation matrix between factors are either zeros or ones. If the orthogonality restriction is relaxed then there would be three fixed parameters (the diagonal elements of this correlation matrix), and three free parameters (the off-diagonal elements, i.e. the inter-factor correlations). Intermediates between these two extremes might also be considered. Such *confirmatory factor analysis models* are essentially part of the general covariance structure and structural equation models to be discussed in

Chapter 14, and so we shall leave details of fitting the models and testing their fit until then.

13.10   Factor Analysis and Principal Component Analysis Compared

Factor analysis, like principal components analysis, is an attempt to explain a set of data in a smaller number of dimensions than one starts with, but the procedures used to achieve this goal are essentially quite different in the two methods. Factor analysis, unlike principal components analysis, begins with a hypothesis about the covariance (or correlational) structure of the variates. Formally, this hypothesis is that a covariance matrix Σ of order and rank p, can be partitioned into two matrices $\Lambda\Lambda'$ and Ψ. The first is of order p but rank k (the number of common factors), whose off-diagonal elements are equal to those of Σ. The second is a diagonal matrix of full rank p, whose elements when added to the diagonal elements of $\Lambda\Lambda'$ give the diagonal elements of Σ. In other words, the hypothesis is that a set of k latent variables exists ($k < p$), and these are adequate to account for the interrelationships of the variates, though not for their full variances. Principal components analysis, on the other hand, is merely a transformation of the data, and no assumptions are made about the form of the covariance matrix from which the data arise. This type of analysis has no part corresponding to the specific variates of factor analysis. Consequently, if the factor model holds but the specific variances are small we would expect both forms of analysis to give similar results. However, if the specific variances are large they will be absorbed into all the principal components, both retained and rejected, whereas factor analysis makes special provision for them. Factor analysis also has the advantage that there is a simple relationship between the results obtained by analysing a covariance matrix and those obtained from a correlation matrix. It should be remembered that principal components analysis and factor analysis are similar in one other respect, namely that they are both pointless if the observed variables are almost uncorrelated – factor analysis because it has nothing to explain and principal components analysis because it would simply lead to components which are similar to the original variables.

13.11   Summary

Factor analysis has probably attracted more critical comment than any other statistical technique. Because the factor loadings are not determined uniquely by the basic factor model many statisticians have complained that investigators can choose to rotate factors in such a way as to get the answer they are looking for. Indeed, Blackith and Reyment (1971) suggest that the method has persisted precisely because it allows the experimenter to impose his preconceived ideas on the raw data. Other criticisms have been concerned with the acceptability of the concept of underlying, unobservable variables; in psychology such a concept may be reasonable, but in other areas far less appealing. Hills (1977) has gone so far as to suggest that factor analysis is not worth the time necessary to understand it and carry it out, and Chatfield and Collins (1980) recommend that factor analysis should not be used in most practical situations. However, we feel that these criticisms are, on the whole, too extreme. Factor analysis

Table 13.8

1	2	3	4	5	6
1·000	0·439	0·410	0·288	0·329	0·248
	1·000	0·351	0·354	0·320	0·329
		1·000	0·164	0·190	0·181
			1·000	0·595	0·470
				1·000	0·464
					1·000

Table 13.9

	Factor loadings		
Subject	F_1	F_2	Communality
1 French	0·553	0·429	0·490
2 English	0·568	0·288	0·406
3 History	0·392	0·450	0·356
4 Arithmetic	0·740	−0·273	0·623
5 Algebra	0·724	−0·211	0·569
6 Geometry	0·595	−0·132	0·372

should be regarded as simply an additional tool for investigating the structure of multivariate observations. The main danger in its use is taking the model too seriously, since it is likely to be only a very idealised approximation to the truth in the situations in which it is generally applied. Such an approximation may, however, prove a valuable starting point for further investigations.

Exercises

13.1 Explain the similarities and differences between factor analysis and principal components analysis.

13.2 The matrix shown in Table 13.8 gives the correlation coefficients between the scores for a sample of 220 boys on six school subjects, namely French, English, History, Arithmetic, Algebra and Geometry (Lawley and Maxwell 1971). A maximum likelihood factor analysis gives the two-factor solution shown in Table 13.9. By plotting these loadings find an orthogonal rotation which makes the solution easier to interpret.

13.3 Find the exact one factor solution for the correlation matrix

$$\mathbf{R} = \begin{pmatrix} 1 & & \\ \frac{1}{3} & 1 & \\ \frac{1}{3} & \frac{1}{10} & 1 \end{pmatrix}$$

Is the solution acceptable?

13.4 The matrix given in Table 13.10 shows the correlation between five socio-economic variables relating to father-son status. Find the residual

Table 13.10

	Variable				
	1	2	3	4	5
1 Father's occ. level	1·00				
2 Father's occ. status	0·52	1·00			
3 Son's educational level	0·45	0·44	1·00		
4 Son's first job status	0·33	0·42	0·54	1·00	
5 Son's later job status	0·33	0·41	0·60	0·54	1·00

Table 13.11

F_1	F_2
0·789	−0·403
0·834	−0·234
0·740	−0·034
0·586	−0·185
0·676	−0·248
0·654	0·140
0·641	0·234
0·629	0·351
0·564	0·054
0·808	0·414

correlations after fitting (a) three principal components, and (b) a two factor model using maximum likelihood. Which approach best explains the observed correlations?

13.5 A two-factor maximum likelihood solution for a particular set of data yielded the estimates of loadings shown in Table 13.11. By making a plot of the loadings find the angle corresponding to the minimum rotation of axes necessary to remove those that are negative. Find the corresponding rotation matrix (see Chapter 2), and the rotated loadings.

The rotation matrix for a varimax rotation is

$$\mathbf{H} = \begin{pmatrix} 0.7278 & -0.6858 \\ 0.6858 & 0.7278 \end{pmatrix}$$

To what angle of rotation does this correspond?

14

Covariance Structure Models

14.1 Introduction

In the models considered in the previous chapter the observed variables were considered to be linear functions of a number of unobservable latent variables. In this chapter these ideas will be extended to situations involving both response and explanatory latent variables, linked by a series of linear equations. Again models will be considered which attempt to explain the statistical properties, generally covariances and correlations, of the measured variables, in terms of the hypothetical latent variables. As with factor analyses models, the primary statistical problem is one of estimating the parameters of the model and determining the goodness of fit.

Such models represent the convergence of relatively independent research traditions in psychometrics, econometrics and biometrics. The idea of latent variables in psychometrics arises from Spearman's early work in general intelligence; the concept of simultaneous directional influences of some variables on others has been part of economics for several decades and the resulting simultaneous equation models have been used extensively by economists but essentially only with observed variables. Path analysis was introduced by Wright (1934) in a biometrics context and later taken up by sociologists such as Blalock (1961, 1963) and Duncan (1969) who demonstrated the value of combining path analytic representations with simultaneous equation models. Finally in the 1970s several workers, most prominent of whom were Jöreskog (1973), Bentler (1980), Browne (1974) and Keesling (1972) combined these procedures into a general method which could in principle deal with extremely complex models in a routine manner. These most recent developments are, according to Cliff (1983), 'the most important and influential statistical revolution to have occurred in the social sciences.'

14.2 A Simple Covariance Structure Model

To introduce a number of general concepts and problems associated with covariance structure models we shall in this section consider a simple example in which there are two latent variables, u and v, three manifest or observed variables, x, x', and y and where the relationships between the variables are as follows:

$$x = u + \delta, \tag{14.1}$$

$$y = v + \epsilon, \tag{14.2}$$

$$x' = u + \delta'. \tag{14.3}$$

If we assume that δ, δ' and ϵ have zero expected values, that δ and δ' are uncorrelated with each other and with u, and that ϵ is uncorrelated with v, then the covariance matrix of the three manifest variables, x, x' and y may be expressed in terms of parameters representing the variances and covariances of the errors of measurement and of the latent variables, giving

$$\Sigma = \begin{pmatrix} \text{Var}(y) & & \\ \text{Cov}(x,y) & \text{Var}(x) & \\ \text{Cov}(x',y) & \text{Cov}(x',x) & \text{Var}(x') \end{pmatrix} \tag{14.4}$$

$$= \begin{pmatrix} \theta_1 + \theta_2 & & \\ \theta_3 & \theta_4 + \theta_5 & \\ \theta_3 & \theta_4 & \theta_4 + \theta_6 \end{pmatrix} \tag{14.5}$$

where $\theta_1 = \text{Var}(v)$, $\theta_2 = \text{Var}(\epsilon)$, $\theta_3 = \text{Cov}(v,u)$, $\theta_4 = \text{Var}(u)$, $\theta_5 = \text{Var}(\delta)$, and $\theta_6 = \text{Var}(\delta')$.

Interest would now centre on determining estimates of the parameters $\theta_1, \theta_2, \ldots, \theta_6$ from the variances and covariances of the observed variables. For this example a difficulty becomes immediately apparent; θ_1 and θ_2 are not uniquely determined. One can be increased by some amount and the other decreased by the same amount without altering Σ. Consequently, different sets of parameter values lead to the same values for the elements of Σ. This problem of the *identification* of parameters will be taken up in more detail later in the chapter.

14.3   Covariance Structure Models: Some Alternative Formulations

The latent variable models in which we are interested imply that the covariance matrix of the manifest variables has a particular form which in the most general terms we may write as

$$\sigma_{jk} = f_{jk}(\boldsymbol{\theta}) \tag{14.6}$$

where σ_{jk} are the elements of the covariance matrix of the observed variables, $\boldsymbol{\theta}' = [\theta_1, \ldots, \theta_t]$ contains the parameters of the model and $f_{jk}(\boldsymbol{\theta})$ is some particular function of these parameter values. Such a model could in principle be fitted by minimising a suitable function measuring the differences between the elements of the observed covariance matrix and those predicted by the model.

Such a general approach is, however, unlikely to be particularly helpful in encouraging the routine application of these models and so a number of less general, but practically more useful procedures have been developed. Two of these have gained widespread acceptance. The first, developed independently by a number of workers including Jöreskog (1973), Keesling (1972) and Wiley (1973), consists essentially of two parts, the *measurement model* which specifies how the manifest variables are related to the latent variables, and the *structural equation model* which specifies how the two types of latent variables (response and explanatory) are linked.

The second formulation is that due to Bentler (1976) in which variables, both manifest and latent, are divided into *dependent* and *independent* and related by a series of equations. Each of these approaches will now be examined in detail.

14.3.1 The LISREL Formulation of Covariance Structure Models

To introduce this approach let us consider a hypothetical model in which there is a single latent response variable η_1 and two latent explanatory variables ξ_1 and ξ_2 related as follows:

$$\eta_1 = \gamma_{11}\xi_1 + \gamma_{12}\xi_2 + \zeta_1 \tag{14.7}$$

where ζ_1 is a random disturbance term. (We shall assume that the means of η_1, ξ_1, ξ_2 and ζ_1 are all zero.) Equation (14.7) is seen to have the same form as the linear regression models discussed in Chapter 8. Here, however, it is known as the *structural model*, and the parameters γ_{11} and γ_{12} can be considered to represent relatively invariant effects in the underlying process generating the relationship between the observed variables.

The latent variables η_1, ξ_1 and ξ_2 cannot be observed directly. Instead measurements are taken on say four variables, x_1, x_2, x_3, and x_4, thought to be indicators of ξ_1 and ξ_2 and on say two variables, y_1 and y_2, thought to be indicators of η_1. The observed variables are assumed to be related to the latent variables in the following way:

$$\begin{aligned}
y_1 &= \lambda_{11}^y \eta_1 + \epsilon_1, \\
y_2 &= \lambda_{21}^y \eta_1 + \epsilon_2, \\
x_1 &= \lambda_{11}^x \xi_1 + \lambda_{12}^x \xi_2 + \delta_1, \\
x_2 &= \lambda_{21}^x \xi_1 + \lambda_{22}^x \xi_2 + \delta_2, \\
x_3 &= \lambda_{31}^x \xi_1 + \lambda_{32}^x \xi_2 + \delta_3, \\
x_4 &= \lambda_{41}^x \xi_1 + \lambda_{42}^x \xi_2 + \delta_4.
\end{aligned} \tag{14.8}$$

The relationships can be written more conveniently in terms of matrices and vectors as

$$\mathbf{y} = \mathbf{\Lambda_y}\boldsymbol{\eta} + \boldsymbol{\epsilon} \tag{14.9}$$

and

$$\mathbf{x} = \mathbf{\Lambda_x}\boldsymbol{\xi} + \boldsymbol{\delta} \tag{14.10}$$

where $\mathbf{y}' = [y_1, y_2]$, $\mathbf{x}' = [x_1, x_2, x_3, x_4]$ $\boldsymbol{\eta}' = [\eta_1]$ $\boldsymbol{\xi}' = [\xi_1, \xi_2]$ $\boldsymbol{\epsilon}' = [\epsilon_1, \epsilon_2]$, $\boldsymbol{\delta}' = [\delta_1, \delta_2, \delta_3, \delta_4]$ and

$$\mathbf{\Lambda}_y = \begin{pmatrix} \lambda_{11}^y \\ \lambda_{21}^y \end{pmatrix}$$

$$\mathbf{\Lambda}_x = \begin{pmatrix} \lambda_{11}^x & \lambda_{12}^x \\ \lambda_{21}^x & \lambda_{22}^x \\ \lambda_{31}^x & \lambda_{32}^x \\ \lambda_{41}^x & \lambda_{42}^x \end{pmatrix}$$

Equations (14.9) and (14.10) relating the observed variables to the appropriate latent variables are seen to be of exactly the same form as the factor analysis model of the previous chapter (see equation (13.2)). These equations constitute the *measurement model*.

Diagrams representing the structural model, the measurement model, and the complete model are shown in Figure 14.1. These diagrams are known as *path diagrams*, and are extremely useful aids in visualising complex models. (Such diagrams and the related concept of *path analysis* were introduced by Wright, 1934.)

In general we shall assume that we have a set of latent response or dependent variables $\boldsymbol{\eta}' = [\eta_1, \eta_2, \ldots, \eta_r]$ and a set of latent explanatory variables $\boldsymbol{\xi}' = [\xi_1, \ldots, \xi_s]$, and that these are related by the following system of linear structural equations:

$$\boldsymbol{\eta} = \mathbf{B}\boldsymbol{\eta} + \boldsymbol{\Gamma}\boldsymbol{\xi} + \boldsymbol{\zeta} \tag{14.11}$$

where $\mathbf{B}(r \times r)$ and $\boldsymbol{\Gamma}(r \times s)$ are coefficient matrices and $\boldsymbol{\zeta}' = [\zeta_1, \zeta_2, \ldots, \zeta_r]$ is a vector of random disturbance terms. We shall assume that the means of $\boldsymbol{\eta}$, $\boldsymbol{\xi}$, and $\boldsymbol{\zeta}$ are zero. The elements of $\mathbf{B}$ represent direct influences of response variables on other response variables and the elements of $\boldsymbol{\Gamma}$ represent direct influences of explanatory variables on response variables.

As an illustration of the structural equations given in (14.11) consider the path diagram shown in Figure 14.2. Here the complete form of 14.11 would be

$$\begin{pmatrix} \eta_1 \\ \eta_2 \\ \eta_3 \end{pmatrix} = \begin{pmatrix} 0 & 0 & 0 \\ \beta_{21} & 0 & 0 \\ \beta_{31} & \beta_{32} & 0 \end{pmatrix} \begin{pmatrix} \eta_1 \\ \eta_2 \\ \eta_3 \end{pmatrix} + \begin{pmatrix} \gamma_{11} & \gamma_{12} & 0 \\ 0 & \gamma_{22} & \gamma_{23} \\ \gamma_{31} & 0 & 0 \end{pmatrix} \begin{pmatrix} \xi_1 \\ \xi_2 \\ \xi_3 \end{pmatrix} + \begin{pmatrix} \zeta_1 \\ \zeta_2 \\ \zeta_3 \end{pmatrix} \tag{14.12}$$

$$\boldsymbol{\eta} = \mathbf{B}\boldsymbol{\eta} + \boldsymbol{\Gamma}\boldsymbol{\xi} + \boldsymbol{\zeta}$$

Measured response variables $\mathbf{y}' = [y_1, y_2, \ldots, y_q]$ and measured explanatory variables $\mathbf{x}' = [x_1, x_2, \ldots, x_p]$ are related to the latent variables $\boldsymbol{\eta}$ and $\boldsymbol{\xi}$ by the equations

$$\mathbf{y} = \boldsymbol{\Lambda}_y \boldsymbol{\eta} + \boldsymbol{\epsilon} \tag{14.13}$$

and

$$\mathbf{x} = \boldsymbol{\Lambda}_x \boldsymbol{\xi} + \boldsymbol{\delta} \tag{14.14}$$

where the elements of $\boldsymbol{\Lambda}_y(q \times r)$ and $\boldsymbol{\Lambda}_x(p \times s)$ are regression coefficients, and $\boldsymbol{\epsilon}$ and $\boldsymbol{\delta}$ are vectors of random disturbance terms. Equations (14.13) and (14.14) are of course directly analogous to the factor analysis models met in the previous chapter. The complete model is known by the acronym LISREL, standing for *linear structural relationships*.

Since $\boldsymbol{\eta}$ and $\boldsymbol{\xi}$ are unobserved they do not have a definite scale. Both the origin and the unit of measurement of each latent variable are arbitrary, and to define the model completely their values for each latent variable must be assigned in some way. The origin has already been taken care of by the assumption that each latent variable has zero mean. Various methods may be employed to fix the scale of the latent variables. The one used in factor analysis is to assume that they are standardised to have unit variance. However, in the programs used for fitting more complex structural equation models it is generally more convenient to define the unit of measurement of the latent variable to be that of one of the observed variables. This may be arranged by

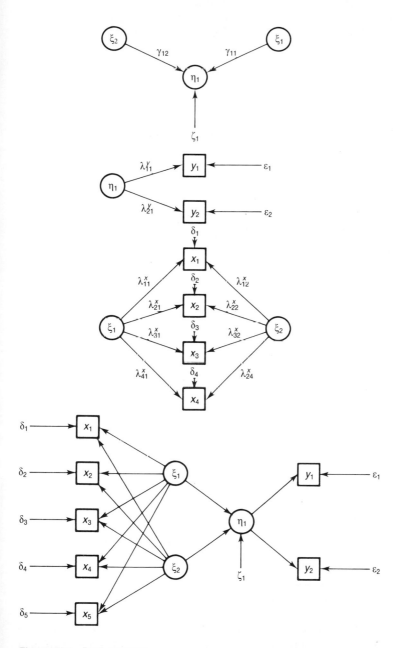

Figure 14.1   Path diagram

simply fixing one of the elements in each column of Λ_x and Λ_y to be unity. The numerical examples given in Section 14.7 should help to clarify this procedure.

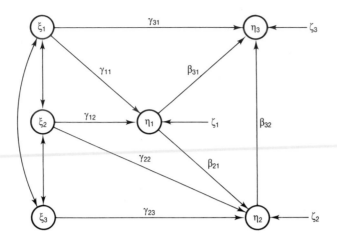

Figure 14.2   Path diagram for model specified by equation (14.12)

If we now make various assumptions about the terms in the structural and measurement parts of the model we can derive the covariance matrix predicted for the observed variables. These assumptions are as follows:

(a) $\boldsymbol{\eta}$ is uncorrelated with $\boldsymbol{\xi}$.
(b) $\mathbf{I} - \mathbf{B}$ is non-singular and $\mathbf{B}$ has zeros in the diagonal.
(c) $\boldsymbol{\epsilon}$ and $\boldsymbol{\delta}$ and uncorrelated with the latent variables.
(d) $\boldsymbol{\Psi}(r \times r)$ is the covariance matrix of $\boldsymbol{\zeta}$.
(e) $\boldsymbol{\Phi}(s \times s)$ is the covariance matrix of $\boldsymbol{\xi}$.
(f) $\boldsymbol{\theta}_\epsilon$ is the covariance matrix of $\boldsymbol{\epsilon}$.
(g) $\boldsymbol{\theta}_\delta$ is the covariance matrix of $\boldsymbol{\delta}$.

From these assumptions it follows that the covariance matrix $\boldsymbol{\Sigma}[(p+q) \times (p+q)]$ of the observed variables is given by

$$\boldsymbol{\Sigma} = \begin{pmatrix} \boldsymbol{\Lambda}_y(\mathbf{I} - \mathbf{B})^{-1}(\boldsymbol{\Gamma}\boldsymbol{\Phi}\boldsymbol{\Gamma}' + \boldsymbol{\Psi})(\mathbf{I} - \mathbf{B})'^{-1}\boldsymbol{\Lambda}_y' + \boldsymbol{\theta}_\epsilon & \boldsymbol{\Lambda}_y(\mathbf{I} - \mathbf{B})^{-1}\boldsymbol{\Gamma}\boldsymbol{\Phi}\boldsymbol{\Lambda}_x' \\ \boldsymbol{\Lambda}_x\boldsymbol{\Phi}\boldsymbol{\Gamma}'(\mathbf{I} - \mathbf{B})'^{-1}\boldsymbol{\Lambda}_y' & \boldsymbol{\Lambda}_x\boldsymbol{\Phi}\boldsymbol{\Lambda}_x' + \boldsymbol{\theta}_\delta \end{pmatrix}$$
$$(14.15)$$

The elements of $\boldsymbol{\Sigma}$ are functions of the elements of $\boldsymbol{\Lambda}_y$, $\boldsymbol{\Lambda}_x$, $\mathbf{B}$, $\boldsymbol{\Gamma}$, $\boldsymbol{\Phi}$, $\boldsymbol{\Psi}$, $\boldsymbol{\theta}_\epsilon$, and $\boldsymbol{\theta}_\delta$. In a particular application some of these elements will be fixed and equal to previously assigned values (often zero or unity); others will be unknown but constrained to be equal to one or more other parameters, and the remainder will be free parameters that are unknown and not constrained to be equal to any other parameter.

Estimation of the parameters in the model will now involve the minimisation of some suitable function of the discrepancies between the observed variances and covariances and those implied by the model. Estimation will be considered in detail in Section 14.5.

14.3.2 The EQS Formulation of Covariance Structure Models

The key concept in this approach to covariance structure models involves the designation of each and every variable in a model as either an *independent* or a *dependent* variable. Dependent variables are those that can be expressed as a structural regression function of other variables; independent variables are never structurally regressed on other variables. Dependent variables are collected in the column vector $\boldsymbol{\eta}$, while independent variables are collected in the vector $\boldsymbol{\xi}$. Thus there are as many elements in $\boldsymbol{\eta}$ as there are dependent variables (both observed and manifest) and as many elements in $\boldsymbol{\xi}$ as there are independent variables (again both observed and manifest). The two types of variable (independent and dependent) are related by a set of structural equations given by

$$\boldsymbol{\eta} = \boldsymbol{\beta}\boldsymbol{\eta} + \boldsymbol{\gamma}\boldsymbol{\xi} \tag{14.16}$$

where $\boldsymbol{\beta}$ contains the coefficients for the regression of $\boldsymbol{\eta}$ variables on each other and $\boldsymbol{\gamma}$ the coefficients for the regression of dependent on independent variables. The parameters of this model are $\boldsymbol{\beta}$, $\boldsymbol{\gamma}$ and the variances and covariances of the independent variables which we represent by the matrix $\boldsymbol{\Phi}$.

It is now relatively easy to show (see Bentler 1989, for details), that the covariance matrices of the observed dependent and independent variables are given by

$$\boldsymbol{\Sigma}_{yy} = \mathbf{G}_y(\mathbf{I} - \boldsymbol{\beta})^{-1}\boldsymbol{\gamma}\boldsymbol{\Phi}\boldsymbol{\gamma}'(\mathbf{I} - \boldsymbol{\beta})^{-1'}\mathbf{G}_y'$$
$$\boldsymbol{\Sigma}_{yx} = \mathbf{G}_y(\mathbf{I} - \boldsymbol{\beta})^{-1}\boldsymbol{\gamma}\boldsymbol{\Phi}\mathbf{G}_x' \tag{14.17}$$
$$\boldsymbol{\Sigma}_{xx} = \mathbf{G}_x\boldsymbol{\Phi}\mathbf{G}_x'$$

where $\mathbf{G}_y$ and $\mathbf{G}_x$ are *selection* matrices which simply represent a way of selecting the observed independent and dependent variables from $\boldsymbol{\xi}$ and $\boldsymbol{\eta}$. For example, if the first six of eight dependent variables in a model are the observed variables, then $\mathbf{G}_y$ is a 6×8 partitioned matrix of the form $(\mathbf{I}, \mathbf{0})$ where $\mathbf{I}$ is a 6×6 identity matrix and $\mathbf{0}$ is a 6×2 null matrix.

Here the acronym for the model is EQS.

14.4 Identification of Parameters in Covariance Structure Models

Before an attempt can be made to estimate the parameters in a covariance structure model, the identification problem must be considered. Essentially this concerns whether a knowledge of the covariance matrix of the observed variables, $\boldsymbol{\Sigma}$ (in terms of its sample estimate, $\mathbf{S}$) allows unique estimates of the unknown parameters in the model. If $\boldsymbol{\Sigma}$ is generated by one and only one set of parameters the model is said to be *identifiable*. More explicitly a model is identified if and only if the following is true for two possible vectors of parameter values, $\boldsymbol{\theta}_1$ and $\boldsymbol{\theta}_2$

if

$$\boldsymbol{\Sigma}(\boldsymbol{\theta}_1) = \boldsymbol{\Sigma}(\boldsymbol{\theta}_2) \tag{14.18}$$

then

$$\boldsymbol{\theta}_1 = \boldsymbol{\theta}_2$$

In the previous chapter we have seen that the parameters in the factor analysis model are not identifiable unless we introduce some constraints, since different sets of factor loadings, corresponding to different rotations of the factor axes, will give rise to the same Σ and in the example in Section 14.2 we saw that θ_1 and θ_2 were not identified, although the sum of the two parameters is identified.

Identifiability depends on the choice of model and on the specification of fixed, constrained and free parameters. If a parameter has the same value in all equivalent structures, the parameter is said to be identified. If all the parameters of a model are identified the whole model is said to be identified. If a parameter is not identified it is not possible to find a consistent estimate of it.

Unfortunately there do not appear to be simple, practicable methods for evaluating identification in the various special cases that might be entertained under the general model. A necessary (but not a sufficient) condition for the identification of all parameters is that

$$t < \frac{1}{2}(p+q+1) \tag{14.19}$$

where t is the number of parameters to be estimated.

In practice it may be necessary to use empirical means to evaluate each particular situation. For example, Jöreskog and Sorbom (1981) suggest that if the information matrix is positive definite it is almost certain that the model is identified. They also demonstrate how the identifiability of a model can be examined in particular cases. Long (1983) gives some further examples of establishing the identifiability of covariance structure models.

14.5   Estimating the Parameters in Covariance Structure Models

Estimates of the parameters in covariance structure models are obtained by minimising some suitable function of the discrepancy between $\mathbf{S}$ and $\Sigma(\theta)$, where $\mathbf{S}$ is the usual unbiased sample covariance matrix obtained from a sample of n observations on the manifest variables, and $\Sigma(\theta)$ is the covariance matrix implied by the model.

The general form of the function minimised is

$$Q = (\mathbf{s} - \boldsymbol{\sigma}(\theta))'\mathbf{W}(\mathbf{s} - \boldsymbol{\sigma}(\theta)) \tag{14.20}$$

where $\mathbf{s}$ is a vector containing the variances and covariances of the observed variables, and $\boldsymbol{\sigma}(\theta)$ a vector containing the corresponding variance and covariances as predicted by the model; the *weight matrix* $\mathbf{W}$ may take a variety of forms depending on the distribution assumed for the manifest variables.

If we assume that the observed variables have a multivariate normal distribution then the fit function, Q, in (14.20) can be shown to take the form

$$Q_N = 2^{-1}\text{trace}[(\mathbf{S} - \Sigma)\mathbf{W}_2]^2 \tag{14.21}$$

and different choices of $\mathbf{W}_2$ lead to three commonly used fit functions:

(1) $\mathbf{W}_2 = \mathbf{I}^{-1}$: ordinary least squares
(2) $\mathbf{W}_2 = \mathbf{S}^{-1}$: generalised least squares
(3) $\mathbf{W}_2 = \Sigma(\hat{\theta})^{-1}$: reweighted least squares.

The last of these is equivalent to maximum likelihood estimation and is more commonly written as involving the minimisation of the function F given by

$$F = \ln |\Sigma(\boldsymbol{\theta})| - \ln |\mathbf{S}| + \text{trace}[\mathbf{S}\Sigma(\boldsymbol{\theta})^{-1}] - (p+q) \tag{14.22}$$

Computer programs based on the LISREL and EQS formulation are available for estimating the parameters in covariance models (see Appendix A) and these will be used to analyse a number of examples in Section 14.7. It is important to bear in mind, however, that the programs are complex and may yield meaningless results even when the user sets up the problem correctly. For example, an identification problem (see Section 14.5) may preclude obtaining a solution, or the initial parameter values supplied may be so far off the optimum that the program may not converge to a final solution, so that some parameter estimates may be completely unreasonable (i.e. negative variances). If some parameters do fall outside their admissible range, either the model is fundamentally wrong or the data are not informative enough.

14.6 Assessing the Fit of a Model

Once a model has been pronounced identified and its parameters estimated, the next problem becomes that of assessing how well it fits the data. One global measure of fit is the likelihood ratio chi-square statistic given by

$$\chi^2 = (N - 1)F_{\min} \tag{14.23}$$

where N is the sample size and $F_{\min}$ is the minimum value of the fitting function in (14.22). If the model is correct and the sample size sufficiently large, this statistic may be used to test the fit of the covariance matrix predicted by the model against the alternative hypothesis that Σ is unconstrained. The degrees of freedom of the statistic are

$$\text{d.f.} = \frac{1}{2}(p+q)(p+q+1) - t \tag{14.24}$$

where t is the number of free parameters in the model.

As a measure of fit, however, the chi-square statistic has limited practical use, since it is a function both of sample size and the closeness of the estimated covariance matrix to the observed covariance matrix. A consequence of this is that the probability of rejecting a model increases as sample size increases even when the matrix of residual covariances, $s_{ij} - \sigma_{ij}(\boldsymbol{\theta})$, contains trivial discrepancies. So in large samples virtually all models would be rejected as statistically untenable. In the context of structural equation modelling with latent variables where, in most practical situations, the exactly correct model is never likely to be known, this is particularly unfortunate since the additional information about a better model contained in the residuals will often be of no practical importance. Perhaps a more reasonable way to use the chi-square statistic is in the comparison of a series of nested models. A large change in chi-square values for two such models compared to the difference in their degrees of freedom may indicate that the additional parameters in one of the models have made a genuine improvement.

Further problems with the chi-square statistic arise when the observations come from a non-normal distribution. Browne (1982) demonstrates that in the case of a distribution with substantial kurtosis the chi-square distribution may be a poor approximation for the likelihood ratio test statistic under the hypothesis that the model is correct. Consequently Browne suggests that before using the test it is advisable to assess the degree of kurtosis, possibly by using Mardia's coefficient of multivariate kurtosis (Mardia, 1970).

Other more satisfactory methods of assessing the fit of a model are (1) visual inspection of residual covariances — they should be small relative to the original covariances; (2) examination of the standard errors of, and the correlation between, parameter estimates — if either is large the model may be almost non-identified; (3) parameter estimates outside their possible range — negative variance or correlations greater than unity are an indication that a model is fundamentally wrong and that it is unsuitable for the data. In addition Jöreskog and Sorbom (1981) describe a number of indices that may be useful in assessing the fit of particular aspects of a model.

Perhaps the most satisfactory approach to assessing fit would be via some form of cross-validation procedure as suggested by Cliff (1983) and Cudeck and Browne (1983). An obvious disadvantage of the technique is that it reduces the sample size by half; in many cases however this disadvantage is greatly outweighed by the indications that the procedure can give of unstable models or those which provide poor predictive validity. In the few situations where it is not possible to cross-validate, Cudeck and Browne suggest using one or other of two indices which provide information similar to cross-validation but are computed from a single sample. The first is the information criterion of Akaike (1973, 1974) and the second is an index proposed by Schwartz (1978). Neither of these indices nor cross-validation are used as often as they should be in practical applications of structural equation modelling.

14.7   Numerical Examples

To begin we shall apply a confirmatory factor analysis model to the data on drug usage rates previously discussed in Chapters 4 and 13. In their detailed discussion of these data Huba *et al.* (1981) suggest a model with three latent variables, with the structure shown in Table 14.1. The corresponding path diagram is shown in Figure 14.3.

Fitting this model results in the parameter estimates shown in Table 14.2; also in this table are the estimated standard errors found for each parameter and the corresponding t statistics for assessing whether or not a parameter differs from zero. All of these statistics are highly significant. The chi-square statistic for the model is 323.9 with 58 degrees of freedom, indicating that the model does not fit the data satisfactorily. Despite this, the matrix of residual correlations (shown in Table 14.3) shows that the majority of residuals are relatively small and that the model might perhaps be reasonably acceptable despite the large chi-square value.

A further model examined by Huba *et al.* allowed correlated error terms for the following pair of drugs: (a) amphetamines and cocaine, (b) tranquillizers and heroin, (c) tranquillisers and amphetamines and (d) drugstore medication and inhalants. The rationale for the inclusion of these extra terms is given in the original paper. The parameter estimates and residual correlations for this model are shown in Table 14.4 and 14.5 respectively. The chi-square

Table 14.1   Structure of confirmatory factor model to be fitted to drug usage data

Drug	F_1	F_2	F_3
1 Cigarettes	X	X	0
2 Beer	X	0	0
3 Wine	X	X	0
4 Liquor	X	0	X
5 Cocaine	0	0	X
6 Tranquillizers	0	0	X
7 Drugstore	0	0	X
8 Heroin	0	0	X
9 Marijuana	0	X	0
10 Hashish	0	X	X
11 Inhalants	0	0	X
12 Hallucinogenics	0	0	X
13 Amphetamines	0	0	X

$$\Phi = \begin{pmatrix} 1\cdot0 & & \\ X & 1\cdot0 & \\ X & X & 1\cdot0 \end{pmatrix}$$

X denotes a free parameter to be estimated.

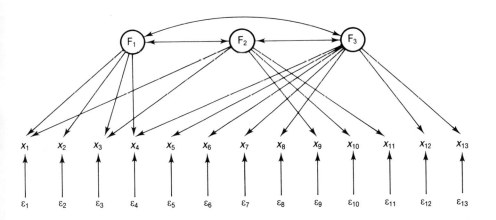

Figure 14.3   Path diagram for model applied to drug usage data

statistic is 209.86 with 54 degrees of freedom. Again this indicates that the fit is not statistically acceptable, although the residuals are again very small. The problem appears to be that with such a large sample size even small and possibly trivial departures from the model are leading to large chi-square values.

Table 14.2   Parameter estimates for confirmatory factor model fitted to drug usage data

Drug	F_1	F_2	F_3
1 Cigarettes	0·36 (10·37)	0·33 (4·40)	0
2 Beer	0·79 (35·02)	0	0
3 Wine	0·88 (23·29)	−0·15 (−4·16)	0
4 Liquor	0·72 (30·67)	0	0·12 (5·44)
5 Cocaine	0	0	0·46 (18·08)
6 Tranquillizers	0	0	0·68 (28·18)
7 Drugstore	0	0	0·36 (13·60)
8 Heroin	0	0	0·48 (18·57)
9 Marijuana	0	0·91 (29·96)	0
10 Hashish	0	0·40 (13·38)	0·38 (13·05)
11 Inhalants	0	0·54 (21·60)	
12 Hallucinogenics	0	0	0·62 (25·23)
13 Amphetamines	0	0	0·76 (32·98)

$$\Phi = \begin{pmatrix} 1\cdot0 & & \\ 0\cdot63\ (23\cdot37) & 1\cdot0 & \\ 0\cdot31\ (10\cdot67) & 0\cdot50\ (18\cdot4) & 1\cdot0 \end{pmatrix}$$

t-values are given in parentheses.

The difference in chi-square values for the two models fitted indicates the improvement in fit made by allowing a number of correlated error terms. The improvement is clearly considerable. Other models in which more correlated error terms are allowed are considered by Huba *et al.* (1981).

Our second example consists of a *simplex* model fitted to a set of data from the Richmond basic skills test (Hieronymous and Lindquist, 1975; France, 1975). Simplex models, first suggested by Guttman (1954) are appropriate when it is considered that there is a flow of influence from one cognitive ability to the next with less complex skills leading on to those which are more complex. Bynner and Romney (1986) examined data collected from 1075 first-year boys and girls (11–12 year olds) in a large secondary school in England; the resulting correlation matrix for 6 of the Richmond sub-tests is shown in Table 14.6. The path diagram for the model of interest is shown in Figure 14.4. The chi-square statistic takes the value 9.5 with 7 degrees of freedom, indicating that the model fits very well. Parameter estimates and standard errors are shown in Table 14.7.

Table 14.3 Residual correlations for model without correlated error terms

	cigs	beer	wine	liquor	cocaine	tranq	drugst	heroin	marij	hash	inhal	hallu	amphet
cigs	0·000												
beer	−0·003	0·000											
wine	0·009	0·002	0·000										
liquor	−0·008	0·002	−0·004	0·000									
cocaine	−0·015	−0·047	−0·039	−0·047	0·000								
tranq	0·15	−0·021	0·005	0·022	0·035	0·000							
drugst	−0·009	0·014	0·039	−0·003	0·042	−0·021	0·000						
heroin	−0·050	−0·055	−0·028	−0·069	0·099	0·033	0·030	0·030					
marij	0·004	−0·012	−0·002	0·009	−0·025	0·008	−0·013	−0·063	0·000				
hash	−0·023	0·025	0·005	0·029	0·034	−0·014	−0·045	−0·057	−0·001	0·000			
inhals	0·094	0·068	0·075	0·065	0·019	−0·044	0·115	0·029	0·054	−0·013	0·000		
hallu	−0·071	−0·065	−0·049	−0·077	−0·009	−0·051	0·010	0·026	−0·077	0·010	0·044	0·000	
amphet	0·033	0·010	0·33	0·026	−0·077	0·030	−0·042	−0·049	0·047	0·025	−0·022	0·039	0·000

Table 14.4 Parameter estimates for second confirmatory factor model applied to drug usage data

Drug	F_1	F_2	F_3
1 Cigarettes	0·36 (10·31)	0·33 (9·40)	0
2 Beer	0·79 (35·02)	0	0
3 Wine	0·88 (23·18)	−0·15 (−4·19)	0
4 Liquor	0·72 (30·84)	0	0·12 (5·26)
5 Cocaine	0	0	0·52 (19·41)
6 Tranquillizers	0	0	0·63 (23·94)
7 Drugstore	0	0	0·33 (12·58)
8 Heroin	0	0	0·46 (17·75)
9 Marijuana	0	0·91 (30·09)	0
10 Hashish	0	0·40 (13·45)	
11 Inhalants	0	0	0·53 (20·81)
12 Hallucinogenics	0	0	0·62 (25·12)
13 Amphetamines	0	0	0·78 (31·57))

$$\Phi = \begin{pmatrix} 1\cdot0 & & \\ 0\cdot54\ (23\cdot56) & 1\cdot0 & \\ 0\cdot31\ (10\cdot40) & 0\cdot50\ (18\cdot3) & 1\cdot0 \end{pmatrix}$$

Covariances amongst error terms
(a) amphetamine–cocaine: −0·13 (−7·24)
(b) heroin–tranquilliser: 0·07 (3·61)
(c) amphetamine–tranquillizer: 0·06 (2·88)
(b) hashish–drugstore: 0·13 (6·19)

Table 14.5 Residual correlations for model allowing some correlated error terms

	cigs	beer	wine	liquor	cocaine	tranq	drugst	heroin	marij	hash	inhal	hallu	amphe
cigs	0·000												
beer	−0·003	0·000											
wine	0·009	0·002	0·000										
liquor	−0·008	0·002	−0·005	0·000									
coc	−0·029	−0·059	−0·480	−0·062	0·000								
tranq	0·032	−0·006	0·018	0·045	0·027	−0·001							
drugst	−0·011	0·012	0·038	−0·005	0·017	−0·009	0·000						
heroin	−0·046	−0·051	−0·025	−0·062	0·080	−0·005	0·028	0·000					
marij	0·004	−0·013	−0·003	0·009	−0·048	0·035	−0·017	−0·056	0·000				
hash	−0·023	0·024	0·005	0·030	0·002	0·011	−0·005	−0·055	−0·000	0·001			
inhal	0·094	0·069	0·076	0·068	−0·011	−0·017	0·107	0·033	0·054	−0·021	0·000		
hallu	−0·070	−0·065	−0·048	−0·074	−0·043	−0·020	0·001	0·030	−0·077	0·001	−0·001	0·000	
amphet	0·035	0·012	0·035	0·032	0·003	0·007	−0·050	−0·041	0·050	0·018	0·025	0·037	0·00

Table 14.6 Correlations between Richmond sub-tests

1	1·00				
2	0·77	1·00			
3	0·62	0·65	1·00		
4	0·56	0·61	0·66	1·00	
5	0·52	0·58	0·63	0·64	1·00
6	0·64	0·69	0·65	0·62	0·61	1·00

Richmond sub-tests
1 Vocabulary
2 Comprehension
3 Spelling,
4 Capital letters,
5 Punctuation,
6 Usage

Table 14.7 Parameter estimates for simplex model fitted to correlations in Table 12.6

Path	Parameter estimates	Standard error	t-value
F_1 to F_2	0·93	0·03	31·95
F_2 to F_6	0·83	0·03	29·46
F_6 to F_3	0·94	0·03	29·08
F_3 to F_4	0·94	0·03	28·45
F_4 to F_5	0·96	0·03	27·12

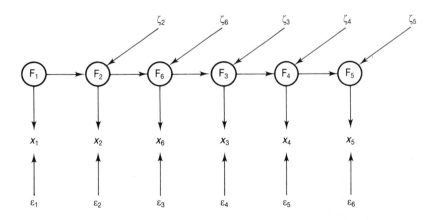

Figure 14.4 Path diagram for simple model applied to Richmond basic skills test data

14.8   Causal Models and Latent Variables – Myths and Realities

During the last decade 'causal analysis' has spread like wildfire across the fields of quantitative social science. Journals of economics, sociology, political science and psychology are aflame with path diagrams, structural equation systems, confirmatory factor analysis and something called LISREL.

Arthur S. Goldberg (1983)

Like many methodological advances before it, the fusion of factor analysis, multiple regression and simultaneous equation models achieved in the 1970s by Jöreskog, Bentler and Browne has been received by many social and behavioural scientists with an enthusiasm which has often overcome their usual critical faculties. In disciplines such as psychology and sociology still seeking the first steps along the road to some type of unifying theory, the suggestion that causal inferences might be demonstrable from correlational data combined with the arrival of a well-documented and forcefully marketed piece of software which makes the application of the procedure relatively simple, has clearly been hard to resist. That old, but still apposite aphorism, 'correlation does not imply causation' often appears to have been conveniently forgotten amongst the mass of path diagrams, parameter estimates and models for everything from theories of intelligence to sexual behaviour. Latent variables are given names and tested as if they had an independent existence and might even be manipulated if necessary. In many cases little attention is given to the purpose that the causal model final settled upon is intended to serve.

All of this is, of course, an extremely unsatisfactory state of affairs, and it is little wonder then that thinly veiled criticisms such as that implicit in the quotation given at the beginning of this section are now beginning to appear. Dealing first with the issue of causality, it is important to recognise that seldom (perhaps never) do structural equation models provide any direct test of their causal assumptions; they are best seen as convenient mathematical fictions which describe the investigator's belief about the causal structure of a set of variables of interest. But however convincing, respectable and reasonable a path diagram and its associated model may appear any causal inferences extracted are rarely more than a form of statistical fantasy. The only satisfactory way to demonstrate causality would be through the active control of variables. As pointed out by Cliff (1983), with correlational data it is simply not possible to isolate the empirical system sufficiently so that the nature of the relaionships among variables can be unambiguously ascertained. Of course, many investigators proposing causal models might argue that they are using the term causal in a purely metaphorical fashion. As pointed out by de Leeuw (1985) such a cavalier attitude towards terminology becomes hard to defend if, for example, educational programs are based on your metaphors, such as the metaphor that 'intelligence is largely genetically determined' or 'allocation of resources to schools has only very minor impact on the careers of students.'

Essentially so-called causal models simply provide a parsimonious description of a set of correlations. This is made explicit in the work of Kuveri and Speed (1982) who demonstrate that such models are equivalent to conditional independence statements. Consequently the use of a package such as LISREL is equivalent to a search amongst conditional independence models for a model with good fit, high explanatory power and which is parsimonious. Such a search is partly guided by objective goodness-of-fit procedure described in the previous section, but also partly by prior knowledge, interpretability etc. Such a combination of objective and subjective procedures makes it difficult to believe that two independent researchers will come up with the same model in any but the simplest of situations. Many of these problems stem of course from the relative lack of well-specified causal theories in the social sciences.

So if the causal in causal modelling is usually a misnomer is the concept of the latent variable more satisfactory? Well in one sense latent variables can never be anything more than is contained in the observed variables and never anything beyond what has been specified in the model. For example, in the statement that verbal ability is whatever certain tests have in common, the empirical meaning is nothing more than a shorthand for the observation of the correlations. It does not mean that verbal ability is a variable that is measurable in any manifest sense. In fact latent variables are essentially hypothetical constructs invented by a scientist for the purpose of understanding some research area of interest, and for which there exists no operational method for direct measurement.

Consequently a question that needs to be asked is can science advance by inferences based upon hypothetical constructs that cannot be measured or empirically tested? According to Lenk (1986) the answer is a resounding – sometimes. For example, atoms in the 18th and 19th centuries were hypothetical constructs which allowed the foundation of thermodynamics; gravity is a further example from physics. Clearly a science *can* advance using the concept of a latent variable, but their importance is not their 'reality' or otherwise but rather to what extent the models of which they are a part are able to describe and predict phenomena (Lakatos, 1977). This point is nicely summarised by Fergusson and Horwood (1986).

> Scientific theories describe the properties of observed variables in terms of abstractions which summarize and make coherent the properties of observed variables. Latent variables, are, in fact, one of this class of abstract statements and the justification for the use of these variables lies not in an appeal to their 'reality' or otherwise but rather to the fact that these variables serve to synthesize and summarize the properties of observed variables.

This point was also made by the participants in the Conference on Systems under Indirect Observation who concluded, after some debate (see Bookstein, 1982), that latent variables are 'as real as their predictive consequences are valid'. Such a comment implies that the justification for postulating latent variables is their theoretical utility rather than their reality.

14.9 Summary

The possibility of making causal inferences about latent variables is one which has great appeal for the social and behavioural scientist simply because many of the concepts in which they are most interested are not directly measurable. Many of the statistical and technical problems in applying the appropriate models to empirical data have largely been solved and sophisticated software such as LISREL means that researchers can investigate and fit extremely complex models routinely. Unfortunately in their rush not to be left behind in the causal modelling stakes many investigators appear to have abandoned completely their proper scientific sceptism, and accepted models as reasonable, simply because it has been possible to fit them to data. This would not be so important if it were not the case that much of the research involved is in areas where action, perhaps far-reaching action, taken on the basis of the findings of the research, can have enormous implications, for example in resources for education, legislation on racial inequality etc. Consequently both producers of such research and audiences or consumers of it need to be particularly concerned that the conclusions reached are valid ones. With this in mind we would like to end with the caveat issued by Cliff (1983):

> ...beautiful computer programs do not really change anything fundamental. Correlational data are still correlational, and no computer program can take account of variables that are not in analysis. Causal relations can only be established through patient, painstaking attention to all the relevant variables, and should involve active manipulation as a final confirmation.

Exercises

14.1 Find the predicted covariance matrix corresponding to a confirmatory factor analysis model with the following structure:

$$\mathbf{\Phi} = \begin{pmatrix} 1.0 & & \\ 0.0 & 1.0 & \\ 0.0 & \phi_{32} & 1.0 \end{pmatrix}$$

$$\mathbf{\Lambda}_x = \begin{pmatrix} \lambda_{11} & 0 & 0 \\ \lambda_{21} & 0 & 0 \\ 0 & \lambda_{32} & 0 \\ 0 & \lambda_{42} & 0 \\ 0 & \lambda_{52} & 0 \\ 0 & \lambda_{62} & 0 \\ 0 & 0 & \lambda_{73} \\ 0 & 0 & \lambda_{83} \end{pmatrix}$$

$$\theta_\delta = \text{diag}(\theta_{ii})$$

Is the model identified?

14.2 The equations given below define a *quasi Markov-simplex* model: draw the corresponding path diagrams and investigate the identification status of the model.

$$y_i = \eta_i + \epsilon_i \qquad i = 1, 2, 3, 4$$
$$\eta_i = \beta_i \eta_{i-1} + \zeta_i \qquad i = 1, 2, 3, 4$$

14.3 The matrix in Table 14.8 shows the correlations obtained between 7 variables in an investigation of subjective probability of victimization (see Smith and Patterson, 1984, for details). The variables y_1, y_2 and y_3 are responses to three questions asking respondents how likely they thought it was that they would be victims of robbery, burglary or vandalism over the next year. The variables x_1 (number of self-reported prior victimizations in the last 12 months), x_2 (respondent's age), x_3 (sex of respondent), and

Table 14.8

y_1	1·0						
y_2	0·575	1·0					
y_3	0·540	0·598	1·0				
x_1	0·169	0·240	0·246	1·0			
x_2	−0·014	−0·144	−0·128	−0·184	1·0		
x_3	−0·023	−0·088	−0·092	−0·148	0·236	1·0	
x_4	0·224	0·215	0·182	0·168	−0·027	−0·102	1·0

x_4 (the rate of personal and property victimizations per 100 households in the resident's neighbourhood), are considered to 'cause' a latent variable, 'perceived risk', the indicators of which are y_1, y_2 and y_3. How well does this model fit the observed correlation matrix?

14.4 An alternative model suggested by Smith and Patterson for the correlation matrix in 14.3, allows an additional path from age to probability of robbery, in the model described above. How can the LISREL formulation given in Section 14.3, be adapted to cope with such a model?

14.5 Give the form of the selection matrices used in EQS for the two models in Exercises 14.3 and 14.4.

References

Agresti, A. (1989), A survey of models for repeated ordered categorical response data. *Statistics in Medicine*, **8**, 1209–24.

Agresti, A. (1990), *Categorical Data Analysis*, John Wiley and Sons, New York.

Aitkin, M. (1978), The analysis of unbalanced cross-classifications, *Journal of the Royal Statistical Society*, A, **141**, 195–223.

Aitkin, M. (1979), A simultaneous test procedure for contingency table models, *Applied Statistics*, **28**, 233–42.

Aitkin, M. (1980), A note on the selection of log-linear models. *Biometrics*, **36**, 173–8.

Aitkin, M. and Clayton, D. (1980), The fitting of exponential, Weibull and extreme value distributions to complex censored survival data using GLIM, *Applied Statistics*, **29**, 156–63.

Aitkin, M., Anderson, D. and Hinde, J. (1981), Statistical modelling of data on teaching styles, *Journal of the Royal Statistical Society*, A, **144**, 419–48.

Aitkin, M., Anderson, D., Francis, B. and Hinde, J. (1989), *Statistical Modelling in GLIM*, Oxford University Press, Oxford.

Akaike, H. (1973), Information theory and an extension of the maximum likelihood principle. In *Second International Symposium on Information Theory* (eds B.N. Petrov and F. Saki), Akademi Koudo, Budapest.

Akaike, H. (1974), A new look at the statistical model identification, *IEEE Trans. Autom. Control*, **19**, 716–23.

Allen, D.M. (1971), Mean square error of prediction as a criterion for selecting variables, *Technometrics*, **13**, 469–75.

Allen, D.M. (1974), The relationship between variable selection and data augmentation and a method for prediction, *Technometrics*, **16**, 125–7.

Allison, P.D. (1984), *Event History Analysis*, Sage University Paper, Sage Publications, Beverly Hills, California.

Anderberg, M.R. (1973), *Cluster Analysis for Applications*, Academic Press, New York.

Anderson, J.A. (1972), Separate sample logistic discrimination. *Biometrika*, **59**, 19–35.

Anderson, R.L., Allen, D.M. and Cady, F. (1972), Selection of predictor variables in multiple linear regression. In *Statistical Papers in Honor of George W. Snedecar* (ed T.A. Bancroft), Iowa State Press, Ames.

Anderson, T.W. (1958), *Introduction to Multivariate Statistical Analysis*, John Wiley and Sons, New York.

Andrews, D.F. (1972), Plots of high dimensional data, *Biometrics*, **28**, 125–36.

Anscombe, F.J. (1973), Graphs in statistical analysis, *The American Statistician*, **27**, 17–21.

Armitage, P. and Berry, G. (1987), *Statistical Methods in Medical Research* (2nd edn), Blackwell Scientific Publications, Oxford.

Baker, R.J. and Nelder, J.A. (1978), *The GLIM System, Release 3: Generalized Linear Interactive Modelling*, Royal Statistical Society, London.

Barnett, V. (ed.) (1981), *Interpreting Multivariate Data*, John Wiley and Sons, London.

Barnett, V. and Lewis, T. (1978), *Outliers in Statistical Data*, John Wiley and Sons, London.

Baum, B.R. (1977), Reductions of dimensionality for heuristic purposes, *Taxon*, **26**, 191–5.

Bennett, N. (1976), *Teaching Styles and Pupil Progress*, Open Books, London.

Bennett, S. and Bowers, D. (1976), *An Introduction to Multivariate Techniques for Social and Behavioural Sciences*, Macmillan, London.

Bentler, P.M. (1976), Multistructure statistical models applied to factor analysis, *Multivariate Behavioral Research*, **11**, 3–25.

Bentler, P.M. (1980), Multivariate analysis with latent variables. Causal Models. *Annual Review of Psychology*, **31**, 419–56.

Bentler, P.M. (1989), *EQS. Structural Equations Manual*, BMDP Statistical Software, Los Angeles, California.

Binder, D.A. (1978), Bayesian cluster analysis, *Biometrika*, **65**, 31–8.

Birch, M.W. (1963), Maximum likelihood in three-way contingency tables, *Journal of the Royal Statistical Society*, B, **25**, 220–33.

Bishop, Y.M.M., Fienberg, S.E. and Holland, P.W. (1975), *Discrete Multivariate Analysis: Theory and Practice*, MIT Press, Cambridge, Mass.

Blackith, R.E. and Reyment, R.A. (1971), *Multivariate Morphometrics*, Academic Press, London.

Blalock, H.M. (1961), Correlation and causality, the multivariate case, *Social Forces*, **39**, 246–51.

Blalock, H.M. (1963), Making causal inferences for unmeasured variables from correlations among indicators, *American Journal of Sociology.*, **69**, 53–62.

Blalock, H.M. (1972), *Social Statistics*, McGraw-Hill, New York.

Blumen, I., Kogan, M. and McCarthy, P.J. (1955), *The Industrial Mobility of Labor as a Probability Process*, Cornell University Press, Ithaca, New York.

Bookstein, F.L. (1982), Panel discussion—modelling and method. In *Systems Under Indirect Observation: Causality Structure and Prediction* (eds K. Joreskog and H. Wold), North Holland, Amsterdam.

Box, G.E.P. (1950), Problems in the analysis of growth and wear curves, *Biometrics*, **6**, 326–89.

Box, G.E.P. (1965), Mathematical models for adaptive control and optimization, *AI. Ch.E. - I. Chem E. Symp.* Series 4, 61.

Box, M.J., Davies, D. and Swann, W. H. (1969), *Non-linear Optimization Techniques*, Oliver and Boyd, Edinburgh.

Breiman, L., Friedman, J.H., Olshen, R.A., and Store, C.J. (1984), *Classification and Regression Trees*, Wadsworth, Belmont.

Browne, M.W. (1974), Generalised least squares estimates in the analysis of covariance structures, *South African Statistical Journal*, **8**, 1–24.

Browne, M.W. (1982), Covariance structures. In *Topics in Applied Multivariate Analysis* (ed. D.M.Hawkins), Cambridge University Press, Cambridge.

Bruntz, S.M., Cleveland, W.S., Kleiner, B. and Warner, J.L. (1974), The dependence of ambient ozone on solar radiation, wind temperature and mixing height, *Proceedings of a Symposium on Atmospheric Diffusion and Air Pollution—American Meteorological Society*, 125–8.

Bryne, P.J. and Arnold, S.F. (1983), Inference about multivariate means for a nonstationery autoregressive model. *Journal of the American Statistical Association*, **78**, 850–5.

Burton, M. (1972), Semantic dimensions of occupation names. In *Multidimensional Scaling*, Vol. 11 (eds A.K. Romney, R.N. Shepard and S.B. Nerlove), Seminar Press, New York.

Buss, A.H. and Durkee, A. (1957), An inventory for assessing different kinds of hostility, *Journal of Consulting Psychology*, **21**, 343–9.

Butler, D. and Stokes, D. (1975), *Political Change in Britain* (2nd edn), Macmillan, London.

Byrne, P.J. and Arnold, S.T. (1983), Inference about multivariate means for a nonstationery autoregressive model. *Journal of the American Statistical Association*, **78**, 850–5.

Bynner, J.M., and Romney, D.M. (1986), Intelligence, fact or artefact: alternative structures for cognitive abilities. *British Journal of Educational Psychology*, **56**, 13–623.

Carroll, J.D. and Chang, J.J. (1970), Analysis of individual differences in multidimensional scaling via an N-way generalization of Eckart–Young decompositon, *Psychometrika*, **35**, 283–319.

Cattell, R.B. (1965), Factor analysis: an introduction to essentials, *Biometrics*, **21**, 190–215.

Chatfield, C. (1982), In discussion of Ramsay (1982).

Chatfield, C. (1985), The initial examination of data. *Journal of the Royal Statistical Society*, A, **148**, 214–53.

Chatfield, C. (1988), *Problem Solving, A Statisticians Guide*, Chapman and Hall, London.

Chatfield, C. and Collins, A.J. (1980), *Introduction to Multivariate Analysis*, Chapman and Hall, London.

Chatterjee, S. and Price, B. (1977), *Regression Analysis by Example*, John Wiley and Sons, New York.

Chernoff, H. (1973), Using faces to representing points in k-dimensional space graphically, *Journal American Statistical Association*, **68**, 361–8.

Child, D. (1970), *The Essentials of Factor Analysis*, Holt, Rinehart Winston, London.

Cliff, N. (1983), Some cautions concerning the application of causal modelling methods, *Multivariate Behavioural Research*, **18**, 115–26.

Cohen, A., Gnanadesikan, R., Kettenring, J.R. and Landwehr, J.M. (1977), Methodological developments in some applications of clustering. In *Proceedings of Symposium on Applications of Statistics* (ed. P.R. Krishnaiah) North Holland, Amsterdam.

Cohn, L.H., Mudge, G.H., Pratter, F. and Collins, J.J. (1981) Five-to- eight-year follow-up of patients undergoing porcine bioprosthetic valve replacement, *New England Journal of Medicine*, **304**, 258–62.

Constantine, A.G. and Gower, J.C. (1978), Graphical representation of asymmetric matrices, *Applied Statistics*, **27**, 297–304.

Cook, R.D (1977), Detection of influential observations in linear regression *Technometrics*, **19**, 15–18.

Cook, R.D. and Weisberg, S. (1982), *Residuals and Influence in Regression*, Chapman and Hall, London.

Coombs, C.H. (1964), *A Theory of Data*, John Wiley and Sons, New York.

Cormack, R.M. (1971), A review of classification, *Journal of the Royal Statistical Society*, A, **134**, 321–67.

Cox, D.R. (1972), Regression models and life tables, *Journal of the Royal Statistical Society*, B, **34**, 187–220.

Cox, D.R. and Oakes, D. (1984), *Analysis of Survival Data*, Chapman and Hall, London.

Cox, D.R. and Snell, E.J. (1989), *Analysis of Binary Data*, (2nd edn), Chapman and Hall, London.

Crowder, M.J. and Hand, D.J. (1990), *Analysis of Repeated Measures*, Chapman and Hall, London.

Cudeck, R., and Browne, M.W. (1983), Cross-validation of covariance structures, *Multivariate Behavioural Research*, **18**, 147–67.

Cunningham, K.M. and Ogilvie, J.C. (1972), Evaluations of hierarchical grouping techniques: a preliminary study, *Computer Journal*, **15**, 209–13.

Davison, M.L. (1983), *Multidimensional Scaling*, John Wiley and Sons, New York.

Day, N.E. (1969), Estimating the components of a mixture of normal distributions, *Biometrika*, **56**, 463–74.

Day, N.E. and Kerridge, D.R. (1967), A general maximum likelihood discriminant, *Biometrics*, **23**, 313–23.

De Leeuw, J. (1985), Book review, *Psychometrika*, **50**, 371–5.

Draper, N.R. and Smith, H. (1981), *Applied Regression Analysis* (2nd edn,), John Wiley and Sons, New York.

Dubes, R. and Jain, A.K. (1979), Validity studies in clustering methodologies, *Pattern Recognition Journal*, **11**, 235–54.

Duda, R. and Hart, P. (1973), *Pattern Classification and Scene Analysis*, John Wiley and Sons, New York.

Duncan, O.D. (1969), Some linear models for two-wave, two variable panel analysis, *Psychological Bulletin*, **72**, 177–82.

Dunn, G. (1981), The role of linear models in psychiatric epidemiology, *Psychological Medicine*, **11**, 179–84.

Dunn, G. (1989), *Design and Analysis of Reliability Studies: Statistical Evaluation of Measurement Errors*, Edward Arnold, London.

Dunn, G. and Clark, P. (1982), Unpublished data.

Dunn, G. and Everitt, B.S. (1982), *An. Introduction to Mathematical Taxonomy*, Cambridge University Press, Cambridge.

Dunn, G. and Master, D. (1982), Latency models: the statistical analysis of response times, *Psychological Medicine*, **12**, 659–65.

Dunn, G. and Skuse, D. (1981), The natural history of depression in general practice: stochastic models, *Psychological Medicine*, **11**, 755–64.

Dunn, M., O'Driscoll, C., Dayson, D., Wills, W. and Leff, J. (1990), The TAPS project 4: an observational study of the social life of long-stay patients, *British Journal of Psychiatry*, 157, 842–848.

Ehrenberg, A.S.C. (1977), Rudiments of numeracy, *Journal of the Royal Statistical Society*, A, **140**, 277–97.

Ekman, G. (1954), Dimensions of colour vision, *Journal of Psychology* **38**, 467–74.

Emerson, J.D. and Soto, M.A. (1983). Transforming Data In *Understanding Robust and Exploratory Data Analysis* (eds D.C. Hoaglin, F. Mosteller and J.W. Tukey), John Wiley and Sons, New York.

Erickson, B.H. and Nosanchuk, T.A. (1979), *Understanding Data*, Open University Press, Milton Keynes.

Everitt, B.S. (1977a), *The Analysis of Contingency Tables*, Chapman and Hall, London.

Everitt, B.S. (1977b), Cluster analysis. In *The Analysis of Survey Data*, Vol. 1, *Exploring Data Structures*, (eds C.A. O'Muircheartaigh and C. Payne), John Wiley and Sons, London.

Everitt, B.S. (1978), *Graphical Techniques for Multivariate Data*, Gower Publications, London.

Everitt, B.S. (1980), *Cluster Analysis* (2nd edn), Gower Publications, London

Everitt, B.S. (1981), A Monte Carlo investigation of the likelihood ratio test for the number of components in a mixture of normal distributions, *Multivariate Behavioural Research*, **16**, 171–80.

Everitt B.S. (1984), *An Introduction to Latent Variable Models*, Chapman and Hall, London.

Everitt, B.S. (1987), *An Introduction to Optimization Methods and Their Application in Statistics*, Chapman and Hall, London.

Everitt, B.S. (1988), A finite mixture model for the clustering of mixed-mode data. *Statistics and Probability Letters*, **6**, 305–9.

Everitt, B.S. (1989), *Statistical Methods for Medical Investigations*, Edward Arnold, London.

Everitt, B.S. and Hand, D.J. (1981), *Finite Mixture Distributions*, Chapman and Hall, London.

Everitt, B.S and Merette, C. (1990), Clustering of Mixed Mode Data. *Journal of Applied Statistics*, **17**, 283–297.

Fergusson, D.M. and Horwood, L.J. (1986), The use and limitations of structural equations models of longitudinal data. Personal Communication.

Feyerabend, P. (1975), *Against Method*, Verso, London.

Fienberg, S.E. (1980), *The Analysis of Cross-Classified Categorical Data* (2nd edn), MIT Press, Cambridge, Mass.

Finn, J.D. (1974), *A General Model for Multivariate Analysis*, Holt, Rinehart and Winston, New York.

Fisher, R.A. (1936), The use of multiple measurements on taxonomic problems, *Annals of Eugenics*, **7**, 179–88.

Fleishman, E.A. and Hempel, W.E. (1954), Changes in factor structure of a complex psychomotor test as a function of practice, *Psychometrika*, **19**, 239–52.

Florek, K., Lukaszewicz, J., Perkal, J., Steinhaus, H., and Zubrzychi, S. (1951) Sur la liason et la division des points d'un ensemble fini, *Colloquium Math.*, **2**, 282–5.

Flury, B. and Riedwyl, H. (1981), Graphical representation of multivariate data by means of asymmetrical faces, *Journal of the American Statistical Association*, **76**, 757–65.

Flury, B. and Riedwyl, H. (1988), *Multivariate Statistics*, Chapman and Hall, London.

France, N. (1975), *Richmond Tests of Basic Skills, Tables of Norms*, Nelson, London.

Francis, I. (1973), Comparison of several analysis of variance programs, *Journal of the American Statistical Association*, **68**, 860–5.

Freka, G.F., Beyts, J.P., Martin, I.M. and Levey, A.B. (1982), unpublished data.

Freireich, E.J., Gehan, E. and Frei, E. (1963), The effect of 6-mercaptopurine on the duration of steroid-induced remissions in acute leukemia: a model for evaluation of other potentially useful therapy, *Blood*, **21**, 699–716.

Friedman, J.H. and Tukey, J.W. (1974), A projection pursuit algorithm for exploratory data analysis, *IEEE Transactions in Computing*, **23**, 881–9.

Gabriel, K.R. (1962). Ante-dependence analysis of an ordered set of variables, *Annals of Mathematical Statistics*, **33**, 201–12.

Gabriel, K.R. (1968). On discrimination using qualitative variables, *Journal of the American Statistical Association*, **63**, 1399–1412.

Gabriel, K.R. (1981), Biplot display of multivariate matrices for inspection of data and diagnosis. In *Interpreting Multivariate Data* (ed. V. Barnett), John Wiley and Sons, London.

Gilbert, E.S. (1968), On discrimination using qualitative variables, *Journal of the American Statistical Association*, **63**, 1399–1412.

Gnanadesikan, R. (1977), *Statistical Data Analysis of Multivariate Observations*, John Wiley and Sons, New York.

Gnanadesikan, R. and Kettenring, J.R.(1989), Discriminant Analysis and Clustering, *Statistical Science*, **4**, 35–69.

Gnanadesikan, R. and Wilk, M.B. (1969), Data analytic methods in multivariate statistical analysis. In *Multivariate Analysis*, Vol. II (ed. P.R. Krishnaiah), Academic Press, New York.

Goldberg, A.S. (1983), Book review. *Contemporary Psychology*, **28**, 858–9.

Goldberg, D.P. (1972), *The Detection of Psychiatric Illness by Questionnaire*, Maudsley Monographs No.21, Oxford University Press, Oxford.

Goldstein, H. (1987). *Multilevel Models in Educational and Social Research* Griffin, London.

Gottman, J.M. (1981), *Times Series Analysis: A Comprehensive Introduction for Social Scientists*, Cambridge University Press, Cambridge.

Gower, J.C. (1966), Some distance properties of latent root and vector methods used in multivariate analysis, *Biometrika*, **53**, 325–38.

Gower, J.C. (1967), Multivariate analysis and multidimensional geometry, *The Statistician*, **17**, 13–25.

Gower, J.C. (1975), Goodness-of-fit criteria for classification and other patterned structures. In *Proceedings of the 8th International Conference on Numerical Taxonomy*, (ed. G. Estabrook), Freeman, London.

Gower, J.C. (1977), The analysis of asymmetry and orthogonality. In *Recent Developments in Statistics* (ed. J. Barra), North-Holland, Amsterdam.-23

Gower, J.C. and Digby, P.G.N. (1981), Expressing complex relationships in two dimensions. In *Interpreting Multivariate Data*, (ed. V. Barnett), John Wiley and Sons, London.

Greenacre, M.J. (1983), Practical correspondence analysis. In *Interpreting Multivariate Data*, (ed. V. Barnett), John Wiley and Sons, London.

Greenacre, M. (1983), *Theory and Applications of Correspondence Analysis*, Academic Press, New York.

Greenhouse, S.W. and Geisser, S. (1959), On methods in the analysis of profile data. *Psychometrika*, **24**, 95–112.

Guttman, L.A. (1954), A new approach to factor analysis: the radex. In *Mathematical Thinking in the Social Sciences* (ed. P.F. Lazarsfeld), Columbia University Press, NewYork.

Hand, D.J. (1981a), Branch and bound in statistical data, *The Statistician*, **30**, 3–16.

Hand, D.J. (1981b), *Discrimination and Classification*, John Wiley and Sons, London.

Hand, D.J. (1982), *Kernel Discriminant Analysis*, John Wiley and Sons, Chichester.

Hand, D.J. (1986) Recent advances in error estimation. *Pattern Recognition Letters*, **4**, 335–46.

Hand, D.J. and Taylor, C.C. (1987), *Multivariate Analysis of Variance and Repeated Measures: A Practical Approach for Behavioural Scientists*, Chapman and Hall, London.

Harshman, R.A. (1972), PARAFAC 2: mathematical and technical notes. In *Working Papers in Phonetics, 22*, University of California, Los Angeles.

Hartigan, J.A. (1975), *Clustering Algorithms*, John Wiley and Sons, New York.

Hartwig, F. and Dearing, B.E. (1979), *Exploratory Data Analysis*, Sage Publications, London.

Healy, M.J.R. (1988), *GLIM: An Introduction*, Oxford University Press, Oxford.

Hieronymous, A.N. and Lindquist, E.P. (1975), *Richmond Tests of Basic Skills. Teacher's Guide*, Nelson, London.

Hills, M. (1977), (Book review), *Applied Statistics*, **26**, 339–40.

Hoerl, A.E. and Kennard, R.W. (1970), Ridge regression: biased estimation for nonorthogonal problems, *Technometrics*, **12**, 55–67.

Hosmer, D.W. (1973), A comparison of iterative maximum likelihood estimates of the parameters of a mixture of two normal distributions under three different types of sample, *Biometrics*, **29**, 761–70.

Hotelling, H. (1933), Analysis of a complex of statistical variables into principal components, *Journal of Educational Psychology*, **24**, 417–41.

Huba, G.J., Wingard, J.A. and Bentler, P.M. (1981), A comparison of two latent variable causal models for adolescent drug use. *Journal of Personality and Social Psychology.*, **40**, 180–93.

Hubert, L.J. (1974), Approximate evaluation techniques for the single-link and complete-link hierarchical clustering procedures, *Journal of the American Statistical Association*, **69**, 698–704.

Jardine, N. and Sibson, R. (1968), The construction of hierarchic and non-hierarchic classifications, *Computer Journal*, **11**, 117–84.

Johnson, S.C. (1967), Hierarchical clustering schemes, *Psychometrika*, **32**, 241–54.

Jolliffe, I.T. (1970), *Redundant variables in multivariate analysis*, Unpublished D.Phil. thesis University of Sussex, England.

Jolliffe, I.T. (1972), Discarding variables in a principal components analysis 1: Artificial Data, *Applied Statistics*, **21**, 160–73.

Jolliffe, I.T. (1973), Discarding variables in a principal component analysis, 2: Real Data, *Applied Statistics*, **22**, 21–31.

Jolliffe, I.T. (1986), *Principal Components Analysis*, Springer, New York.

Jolliffe, I.T. (1989), Rotation of ill-defined principal components, *Applied Statistics*, **38**, 139–48.

Jones, B. and Kenward, M.G. (1989), *Design and Analysis of Cross-Over Trials*, Chapman and Hall, London.

Jones, M.C. and Sibson, R. (1987), What is projection pursuit?, *Journal of the Royal Statistical Society*, A, **150**, 1–36.

Jöreskog, K.G. (1967), Some contributions to maximum likelihood factor analysis, *Psychometrika*, **32**, 443–82.

Jöreskog, K.G. (1973), Analysis of covariance structures. In *Multivariate Analysis* Vol. III (ed. P.R. Krishnaiah), Academic Press, New York.

Jöreskog, K.G. and Lawley, D.N. (1968), New methods in maximum likelihood factor analysis, *British Journal of Mathematical and Stastical Psychology*, **21**, 85–96.

Jöreskog, K.G. and Sörbom, D. (1981), *LISREL V: Analysis of linear structural relationships by maximum likelihood and least squares methods*, Research Report 81-8, Department of Statistics, Uppsala, Sweden.

Kaiser, H.F. (1958), The varimax criterion for analytic rotation in factor anlaysis, *Psychometrika*, **23**, 187–200.

Kalbfleisch, J.D. and Prentice, R.L. (1980), *The Statistical Analysis of Failure Time Data*, John Wiley and Sons, New York.

Kaplan, E.L. and Meier, P. (1958), Nonparametric estimation from incomplete observations, *Journal of the American Statistical Association*, **53**, 457–81.

Kay, R. (1984), Goodness of fit methods for the proportional hazards regression model: a review, *Review Epidemiology et Sante Publication*, **32**, 189–98.

Kendall, D.G. (1975), The recovery of structure from fragmentary information, *Philosophical Transactions of the Royal Society*, (A), **279**, 547–82.

Kendall, M.G. (1975), *Multivariate Analysis*, Griffin, London.

Kendall, M.G. and Stuart, A. (1980), *The Advanced Theory of Statistics*, Griffin, London.

Kenward, M.G. (1987), A method of comparing profiles of repeated measurements. *Applied Statistics*, **36**, 296–308.

Kihlberg, J.K., Narragon, E.A. and Campbell, B.J. (1964), Automobile crash injury in relation to car size, *Cornell Aero.Lab. Report No. VJ-1823- R11.*

Kleiner, B. and Hartigan, J.A. (1981), Representing points in many dimensions by trees and castles, *Journal of the American Statistical Association*, **76**, 260–69.

Koch, G.G., Landis, J.R., Freeman, J.L., Freeman, D.H. and Lehnen, R.G. (1977), A general methodology for the analysis of experiments with repeated measurement of categorical data, *Biometrics*, **33**, 133–58.

Kruskal, J.B. (1964a), Multidimensional scaling by optimizing goodness-of-fit to non-metric hypotheses, *Psychometrika*, 209–115–29.

Kruskal, J.B. (1964b), Non-metric multidimensional scaling: a numerical method, *Psychometrika*, **29**, 115–29.

Kruskal, J.B. and Wish, M. (1978), *Multidimensional Scaling*, Sage Publications, London.

Krzanowki, W.J. (1977), The performance of Fisher's linear discriminant function under non-optimal conditions, *Technometrics*, **19**, 191–200.

Kuveri, H. and Speed, T.P. (1982), Structural analysis of multivariate data: a review. In *Sociological Methodology* (ed. S. Leinhardt), Jossey-Buss, San Francisco.

Laird, N.M. and Olivier, D. (1981), Covariance analysis of censored survival data using log-linear analysis techniques, *Journal of the American Statistical Association*, **76**, 31–240.

Laird, N.M. and Ware, J.H. (1982), Random-effects models for longitudinal data. *Biometrics*, **38**, 963 74.

Lakatos, I. (1977), *The Methodology of Scientific Research Programs*, Cambridge University Press, Cambridge.

Lawley, D.N. (1940), The estimation of factor loadings by the method of maximum likelihood, *Proceedings of the Royal Society of Edinburgh*, **A60**, 64–82.

Lawley, D.N. (1941), Further investigations in factor estimation, *Proceedings of the Royal Society of Edinburgh*, **A61**, 176–85.

Lawley, D.N. (1943), The application of the maximum likelihood method to factor analysis, *British Journal of Psychology*, **33**, 172–5.

Lawley, D.N. (1967), Some new results in maximum likelihood factor analysis, *Proceedings of the Royal Society of Edingburgh*, **A67**, 256–64.

Lawley, D.N. and Maxwell, A.E. (1971), *Factor Analysis as a Statistical Method* (2nd edn), Butterworths, London.

Lazarsfeld, P.L. and Henry, N.W. (1968), *Latent Structure Analysis*, Houghton Mifflin, Boston.

Lee, E.T. (1980), *Statistical Methods for Survival Data Analysis*, Wadsworth, California.

Lenk, P.J. (1986), Book review. *Journal of the American Statistical Association*, 1123–4.

Lindsey, J.K. (1989), *The Analysis of Categorical Data using GLIM*, Springer-Verlag, New York.

Little, R.A. and Rubin, D.B. (1987), *Statistical Analysis with Missing Data*, John Wiley and Sons, New York.

Ling, R.F. (1973), A computer generated aid for cluster analysis, *Communications of the ACM*, **16**, 355–61.

Long, J.S. (1983), *Covariance Structure Models. An introduction to LISREL*, Sage University Paper, 34. Sage Publications, Beverley Hills.

McLachlan, G.J. and Basford, K.E. (1988), *Mixture Models. Inference and Applications to Clustering*, Marcel Dekker, New York.

McCullagh P. (1980), Regression models for ordinal data, *Journal of the Royal Statistical Society*, B, **42**, 109–27.

McCullagh, P. and Nelder, J.A. (1989), *Generalized Linear Models* (2nd edn), Chapman and Hall, London

McDonald, R.P. (1978), A simple comprehensive model for the analysis of covariance structures, *British Journal of Mathematical and Stastical Psychology*, **31**, 59–72.

MacDonnell, W.R. (1902), On criminal anthropometry and the identification of criminals, *Biometrika*, **1**, 177–227.

McQuitty, L.L. (1957), Elementary linkage analysis for isolating orthogonal and oblique types and typal relevances, *Educational Psychology Measurement*, **17**, 207–29.

Mahon, B.H. (1977), Statistics and decisions: the importance of communication and the power of graphical presentation, *Journal of the Royal Statistical Society*, A, **140**, 298–306.

Mallows, C.L. (1973) Some comments on Cm, *Technometrics*, **15**, 661–75.

Malone, L.C. (1984), Contrasting split plot and repeated measures experiments and analyses, *The American Statistician*, **38**, 21–31.

Manley, B.F.J. (1986), *Multivariate Statistical Methods: A Primer*, Chapman and Hall, London.

Mardia, K.V. (1970), Measures of multivariate shewness and kurtosis with applications, *Biometrika*, **57**, 519–30.

Mardia, K.V., Kent, J.T. and Bibby, J.M. (1979), *Multivariate Analysis*, Academic Press, London.

Marriott, F.H.C. (1974), *The Interpretation of Multiple Observations*, Academic Press, London.

Matthews, J.N.S., Altman, D.G., Campbell, M.J. and Royston, P. (1990) Analysis of serial measurements in medical research, *British Medical Journal*, **300**, 230–35.

Maxwell, A.E. (1977), *Multivariate Analysis in Behavioural Research*, Chapman and Hall, London.

Melzak, R. (1975), The McGill pain questionnaire: major properties and scoring methods, *Pain*, **1**, 277–99.

Milligan, G.W. (1981). A Monte Carlo study of thirty internal criterion measures for cluster analysis, *Psychometrika*, **46**, 187–99.

Milligan, G.W. and Cooper, M.C. (1985), An examination of procedures for determining the number of clusters in a data set, *Psychometrika*, **50**, 159–79.

Mojena, R. (1977), Hierarchical grouping methods and stopping rules: an evaluation, *Computer Journal*, **20**, 359–63.

Monlezvn, C.J., Blouin, D.C. and Malone, L.C. (1984), Contrasting split plot and repeated measures experiments and analysis, *The American Statistician*, **38**, 21–31.

Moore, D.H. (1973), Evaluation of five discrimination procedures for binary variables, *Journal of the American Stastical Association*, **68**, 339–404.

Moore, D. (1989), Personal communication.

Morgan, B.J.T. (1981), Three applications of methods of cluster analysis,' *The Statistician*, **30**, 205–24.

Morrison, D.F. (1967), *Multivariate Statistical Methods*, McGraw-Hill, New York.

Mosteller, F. and Tukey, J.W. (1977), *Data Analysis and Regression*, Addison-Wesley, Reading, Mass.

Murray, J.D., Dunn, G., Williams, P. and Tarnopolsky, A. (1981), Factors affecting the consumption of psychotropic drugs, *Psychological Medicine*, **11**, 551–60

Murray, J.D., Dunn, G. and Tarnopolsky, A. (1982), Self-assessment of health: an exploration of the effects of physical and psychological symptoms, *Psychological Medicine*, **12**, 371–8.

Muthen, B. (1981), A general structural equation model with ordered categorical and continuous latent variable indicators, *Research Report 81-9*, Department of Statistics, Uppsala, Sweden.

Nelder, J.A. (1977), A reformulation of linear models, *Journal of the Royal Statistical Society*, A, **140**, 48–63.

Nelder, J.A. and Wedderburn, R.W.M. (1972), Generalized linear models, *Journal of the Royal Statistical Society*, A, **135**, 370–84.

Nie, N.H., Hull, C.H., Jenkins, J.G., Steinbrenner, K. and Bent, D.H. (1975), *SPSS: Statistical Package for the Social Services*, (2nd edn), New York, McGraw-Hill.

Paykel, E.S. and Rassaby, E. (1978), Classification of suicide attempters by cluster analysis, *British Journal of Psychiatry*, **133**, 45–52.

Pearson, K. (1901), On lines and planes of closest fit to systems of points in space, *Philosophical Magazine*, **2**, 559–72.

Powell, G.E., Clark, E. and Bailey, S. (1979), Categories of aphasia a cluster analysis of Schuell test profiles, *British Journal of Disorders of Communications*, **14**, 111–22.

Prentice, R.L. and Gloeckler, L.A. (1978), Regression analysis of grouped survival data with application to breast cancer data, *Biometrics*, **34**, 57–67.

Ramsay, J.O. (1977), Maximum likelihood estimation in multidimensional scaling, *Psychometrika*, **42**, 241–66.

Ramsay, J.O (1978), Confidence regions for multidimensional scaling analysis, *Psychometrika*, **43**, 145–60.

Ramsay, J.O. (1982), Some statistical approaches to multidimensional scaling data, *Journal of the Royal Stastical Society*, A, **145**, 285–312.

Ramsay, J.O. (1983), *Multiscale Manual*, Scientific Software, Chicago.

Rao, C.R. (1965), The theory of least squares when the parameters are stochastic and its application to the analysis of growth curves, *Biometrika*, **52**, 447–58.

Rawlings, J.O. (1988), *Applied Regression Analysis: A Research Tool*, Wadsworth and Brooks/Cole, Belmont, California.

Reading, A.E., Everitt, B.S. and Sledmere, C.M. (1982), The McGill pain questionnaire: a replication of its construction, *British Journal of Clinical Psychology*, **21**, 339–49.

Ries, P.M., and Smith, H. (1963), The use of chi-square for preference testing in multidimensional problems, *Chemical Engineering Progress*, **59**, 39–43.

Rohlf, F.J. (1970), Adaptive hierarchical clustering schemes, *Systematic Zoology*, **19**, 58–82.

Rohlf, F.J. and Fisher, D.L. (1968), Test for hierarchical structure in random data sets, *Systematic Zoology*, **17**, 407–12.

Romney, A.K., Shepard, R.N. and Nerlove, S.B. (1972), *Multidimensional Scaling*, Vols I and II, Seminar Press, New York.

Rothkopf, E.Z. (1957), A measure of Stimulus similarity and errors in some paired associate learning tasks, *Journal of Experimental Psychology*, **53**, 94–101.

Sammon, J.W. (1969), A non-linear mapping for data structure analysis, *IEEE Transactions in Computers*, **C18**, 401–9.

Sattath, S. and Tversky, A. (1977), Additive similarity trees, *Psychometrika*, **42**, 319–45.

Schuell, H. (1965), *Differential Diagnosis of Aphasia*, Minneapolis, University of Minnesota Press.

Searle, S.R. (1971), *Linear Models*, John Wiley and Sons, New York.

Searle, S.R. (1987), *Linear Models for Unbalanced Data*, John Wiley and Sons, New York..

Shepard, R.N. (1962), The analysis of proximities: multidimensional scaling with an unknown distance function, 1, *Psychometrika*, **27**, 219–46.

Siegel, S. (1956), *Nonparametric Statistics*, McGraw-Hill, New York.

Smith, S.P. and Dubes, R. (1980), Stability of a hierarchical clustering, *Pattern Recognition*, **12**, 177–87.

Sneath, P.H.A. (1957), The application of computers to taxonomy, *Journal of Genetic Microbiology*, **17**, 201–26.

Spearman, C. (1904), General Intelligence, objectively determined and measured, *American Journal of Psychology*, **15**, 201–93.

Spence, I. (1970), *Multidimensional scaling: an empirical and theoretical investigation*, Ph.D. thesis, University of Toronto.

Spence, I. (1972), *An aid to the estimation of dimensionality in nonmetric multidimensional scaling*, University of Western Ontario Research Bulletin No. 229.

Spence, I. and Graef, J. (1974), The determination of the underlying dimensionality of an empirically obtained matrix of proximities, *Multivariate Behavioural Research*, **9**, 331–42.

Srole, L., Langner, T.S., Michael, S.T., Opler, M.K., and Rennie, T.A.C. (1962), *Mental Health in the Metropolis: The Midtown Manhattan Study*, McGraw-Hill, New York.

Stanek, E.J. III (1990), A two-step method for understanding and fitting growth curve models, *Statistics in Medicine*, **9**, 841–51.

Tarnopolsky, A. and Morton-Williams, J. (1980), *Aircraft and Prevalence of Psychiatric Disorders*, Social and Community Planning Research, London.

Thurstone, L.L. (1931), Multiple factor analysis, *Psychology Review*, **38**, 406–27.

Titterington, D.M., Smith, A.F.M. and Makov, U.E. (1985), *Statistical Analysis of Finite Mixture Distributions*, John Wiley and Sons, New York.

Tucker, L.R. (1964), The extension of factor analysis to three-dimensional matrices. In *Contributions to Mathematical Psychology* (eds N. Fredriksen and H. Gulliksen), Holt, Rinehart and Winston, New York.

Tucker, L.R. (1972), Relations between multidimensional scaling and three-mode factor analysis, *Psychometrika*, **37**, 3–27.

Tucker L.R. and Messick, S. (1963), An individual differences model for multidimensional scaling, *Psychometrika*, **28**, 333–67.

Tukey, P.A. and Tukey, J.W. (1981a), Preparation, pre-chosen sequence of views. In *Interpreting Multivariate Data* (ed. V. Barnett), John Wiley and Sons, London.

Tukey, P.A. and Tukey, J.W. (1981b), Data-driven view selection; agglomeration and sharpening. In *Interpreting Multivariate Data*, (ed. V. Barnett), John Wiley and Sons, London.

Tukey, P.A. and Tukey, J.W. (1981c), Summarization smooth; supplemented views. In *Interpreting Multivariate Data*, (ed. V. Barnett), John Wiley and Sons, London.

Wagenaar, W.A. and Padmos, P. (1971), Quantitative interpretation of stress in Kruskal's multidimensional scaling technique, *British Journal of Mathematical and Statistical Psychology*, **24**, 101–10.

Ward, J.H. (1963), Hierarchical grouping to optimize an objective function. *Journal of the American Statistical Association*, **58**, 236–44.

Wheaton, B., Muthen, B., Alwin, D. and Summers, G. (1977), Assessing reliability and stability in panel models. In *Sociological Methodology* (ed. D.R. Heise), Jossey Bass, San Fransisco.

Whittaker J. and Aitkin, M. (1978). A flexible strategy for fitting complex log-linear models. *Biometrics*, **34**, 487–95.

Wiley, D.E. (1973). The identification problem for structural equation models with unmeasured variables. In *Structural Equation Modes in the Social Sciences*, (eds A.S. Goldberger and O.D. Duncan), Seminar Press, New York.

Williams, W.T., Lance, G.N., Dale, M.B. and Clifford, H.T. (1971), Controversy concerning the criteria for taxonometric strategies, *Computer Journal*, **14**, 162–5.

Winer, B.J. (1971), *Statistical Principles in Experimental Design*, (2nd edn) McGraw-Hill, New York.

Woese, C.R. (1981), Archaebacteria, *Scientific American*, **244**, 94–106.

Wolfe, J.H. (1970), Pattern clustering by multivariate mixture analysis, *Multivariate Behavioural Research*, **5**, 329–50.

Wolfe, J.H. (1971), A Monte Carlo study of the sampling distribution of the likelihood ratio for mixtures of multinormal distributions (Naval Personnel and Training Research Laboratory), Technical Bulletin, STB 72-2 (San Diego, California, 92152).

Wonnacott, R.J. and Wonnacott. T.H. (1981), *Regression; A Second Course in Statistics*, John Wiley and Sons, New York.

Wright, S. (1934), The method of path coefficients, *Annals of Mathematical Statistics*, **5**, 161–215.

Yule, W., Berger, M., Butler, S., Newham, V. and Tizard, J. (1969), The WPPSI: an empirical evaluation with a British sample,' *British Journal of Educational Psychology*, **39**, 1–13.

Appendix A
Programs and Packages

Most of the methods of analysis that are described in this text have to be performed with the aid of a computer and a suitable program. More and more frequently this means the use of a personal computer (PC) and a statistical software package. In this appendix we give a brief description of the most commonly used of these. We make no attempt to evaluate any of these packages, nor do we claim to have produced an exhaustive list.

A.1   General Purpose Packages

The three best known of these are SAS (Statistical Analysis System), SPSS (Statistical Package for the Social Sciences), and BMDP (Biomedical Computer Programs). Three others are STATGRAPHICS, GENSTAT and MINITAB. These general-purpose packages offer a wide range of statistical techniques including many of those described in this text. They all contain programs for analysis of variance and multiple regression and most of the other methods described in this book will be found in at least one of the packages. All are to a large extent 'user friendly' and have excellent user manuals. One or other of these packages will meet the statistical needs of the majority of users.

A.2   Generalised Linear Modelling

For the more ambitious researcher and for the applied statistician, an attractive package, particularly for the application of log-linear and linear-logistic models, is GLIM. It may also be used for analysis of variance, multiple regression and the analysis of survival data. The program allows the user to specify a range of error distributions and link functions and, together with the ability to store and use user-defined macros (subroutines), this allows an enormous amount of flexibility in modelling. A library of macros is also provided with the package.

GENSTAT is also useful for fitting generalised linear models and non-linear models. Another program which is particularly suitable for the latter is MLP.

Variance-component models can be fitted through programs in SAS and BMDP and the special-purpose program, REML.

EGRET is an easy-to-use program suitable for fitting models to contingency tables and for the analysis of survival data.

A.3   Specialised Programs

Various programs and packages have been developed which are applicable to a particular area of multivariate analysis. For example, the CLUSTAN package allows an extensive range of clustering algorithms to be applied to data sets, and a wide variety of similarity and distance measures to be considered.

MULTISCAL is a package for multidimensional scaling which allows more formal models to be applied, and ALSCAL is a comprehensive scaling package which is now available as part of the SAS system mentioned earlier.

For confirmatory factor analysis and structural equation modelling with latent variables the two packages most widely used are LISREL VII and EQS. With the former, models have to be specified in terms of eight parameter matrices; with the latter these models are specified directly in terms of a set of equations.

A.4   Statistical Languages

Apart from the packages designed to allow the easy application of standard methods, a number of statistical programming languages have been developed that enable the more experienced user to develop individual programs for specialised applications. Mention has already been made of this facility within GLIM. GENSTAT allows for user-defined programs and the easy manipulation of vectors and matrices, as does the matrix-manipulating procedure within SAS. Other examples include S-PLUS, SC and GAUSS. S-PLUS has in addition extensive graphical capabilities.

Appendix B
Answers to Selected Exercises

B.1 Chapter 2

2.1 Matrix A is of rank 5.

2.5

$$f(x) = \frac{1}{\sqrt{(2\pi)}\sigma} \exp -\frac{1}{2}\left(\frac{x-\mu}{\sigma}\right)^2$$

$$\log_e f = -\frac{1}{2}\log_e(2\pi) - \frac{1}{2}\log_e \sigma^2 - \frac{1}{2}\left(\frac{x-\mu}{\sigma}\right)^2$$

$$\frac{\partial \log_e f}{\partial \mu} = \frac{x-\mu}{\sigma^2}. \quad E\left(\frac{\partial \log_e f}{\partial \mu}\right)^2 = \frac{1}{\sigma^2}.$$

$$\frac{\partial \log_e f}{\delta\sigma^2} = -\frac{1}{2\sigma^2} + \frac{1}{2\sigma^4}(x-\mu)^2.$$

$$E\left(\frac{\partial \log_e f}{\partial\sigma^2}\right)^2 = E\left\{\frac{(x-\mu)^4}{4\sigma^8} - \frac{(x-\mu)^2}{2\sigma^6} + \frac{1}{4\sigma^4}\right\}$$

$$= \frac{1}{2\sigma^4}$$

(since for a normal distribution $E(x-\mu)^4 = 3\sigma^4$).

$$E\left(\frac{\partial \log_e f}{\partial\sigma^2}\right)\left(\frac{\partial \log_e f}{\partial\mu}\right) = E\left(\frac{x-\mu}{\sigma^2}\right)\left\{\frac{(x-\mu)^2}{2\sigma^2} - \frac{1}{2\sigma^2}\right\}$$

$$= 0$$

(since for a normal distribution $E(x-\mu)^3 = 0$).
Consequently the information matrix, $\mathbf{I}$, is given by

$$\mathbf{I} = \begin{bmatrix} \frac{1}{\sigma^2} & 0 \\ 0 & \frac{1}{2\sigma^4} \end{bmatrix}$$

and the covariance matrix, $(n\mathbf{I})^{-1}$, would be

$$n\mathbf{I}^{-1} = \begin{bmatrix} \frac{\sigma^2}{n} & 0 \\ 0 & \frac{2\sigma^4}{n} \end{bmatrix}$$

An estimate of this matrix would be obtained by substituting the m.l.e. of σ^2, so that

$$\hat{\mathrm{V}}\mathrm{ar}(x) = s^2/n; \quad \hat{\mathrm{V}}\mathrm{ar}(s^2) = 2s^4/n$$
$$\hat{\mathrm{C}}\mathrm{ov}(\bar{x}, s^2) = 0.$$

2.8

$$f(x) = x^2 + 2x + 3$$
$$\frac{df(x)}{dx} = 2x + 2$$

Setting

$$\frac{df(x)}{dx} = 0$$

gives $x = -1$

$$\frac{d^2 f(x)}{dx^2} = 2$$

so that $x = -1$ corresponds to a minimum value of $f(x)$.

2.11 Variables y_1 and y_2 are linear functions of $\mathbf{x}$ and the result in Section 2.4 is easily extended to show that if

$$y_1 = \mathbf{a}'\mathbf{x} \quad \text{and} \quad y_2 = \mathbf{b}'\mathbf{x}$$

then

$$\mathrm{Cov}(y_1, y_2) = \mathbf{a}'\Sigma\mathbf{b}$$

Here

$$\mathbf{a}' = [1, 1, 1], \quad \mathbf{b}' = [1, 0, -1],$$

so that

$$\mathrm{Cov}(y_1, y_2) = [1 \quad 1 \quad 1] \begin{bmatrix} \sigma_{11} & \sigma_{12} & \sigma_{13} \\ \sigma_{21} & \sigma_{22} & \sigma_{23} \\ \sigma_{31} & \sigma_{32} & \sigma_{33} \end{bmatrix} \begin{bmatrix} 1 \\ 0 \\ -1 \end{bmatrix}$$
$$= \sigma_{11} + \sigma_{21} - \sigma_{23} - \sigma_{33}$$

B.2   Chapter 4

4.1 We first need to find the expected value of $\mathbf{x}$

$$E(\mathbf{x}) = \begin{bmatrix} E(x_1) \\ E(x_2) \end{bmatrix} = \begin{bmatrix} p \\ q \end{bmatrix}$$

The covariance matrix of x is defined to be

$$E(\mathbf{x} - E(\mathbf{x}))(\mathbf{x} - E(\mathbf{x}))'$$
$$= E \begin{bmatrix} x_1 & -p \\ x_2 & -q \end{bmatrix} [x_1 - p \quad x_2 - q]$$
$$= E \begin{bmatrix} (x_1 - p)^2 & (x_1 - p)(x_2 - q) \\ (x_2 - q)(x_1 - p) & (x_2 - q)^2 \end{bmatrix}$$
$$= \begin{bmatrix} p(1 - p) & -pq \\ -pq & q(1 - p) \end{bmatrix}$$

B.3 Chapter 5

5.1 For missing data simply omit, both in the numerator and the denominator of the stress goodness-of-fit measure, the terms which correspond to the missing dissimilarities.

Kruskal (1964a) suggests two approaches to ties. One is to say that when $\delta_{ij} = \delta_{kl}$ we do not care which of d_{ij} and d_{kl} is larger nor whether they are equal or not. The second approach is to say that when $\delta_{ij} = \delta_{kl}$, d_{ij} should be equal to d_{kl}, and to downgrade a configuration if this is not so.

Both approaches can be accommodated by small alterations in the goodness-of-fit criterion, and are described in Kruskal (1964a).

B.4 Chapter 6

6.1 Let x, y and z be any three objects, and suppose that, at fusion level α_j, x and y are in the same cluster, and at fusion level α_k, y and z are in the same cluster. Since the clusters are hierarchical, one of these includes the other. This will be the cluster corresponding to the larger of j and k. Let this be the integer e so that at α_e, x, y and z are all in the same cluster. Then,

$$d(x, z) \leq \alpha_1$$

but since

$$e = \max\{j, k\}$$
$$\alpha_e = \max\{\alpha_j, \alpha_k\}$$

so that

$$d(x, z) \leq \max\{\alpha_j, \alpha_k\}$$

i.e.

$$d(x, z) \leq \max\{d(x, y), d(y, z)\}.$$

6.2 The single linkage distance between cluster k and the cluster formed by the fusion of clusters i and j is defined to be $\min \{d_{ki}, d_{kj}\}$. According to the formula given in Exercise 6.2 this distance, $d_{k(ij)}$ is

$$d_{k(ij)} = \frac{1}{2}d_{ki} + \frac{1}{2}d_{kj} - \frac{1}{2}|d_{ki} - d_{kj}|$$

if

$$d_{ki} > d_{kj}, \quad |d_{ki} - d_{kj}| = d_{ki} - d_{kj}$$

therefore

$$d_{k(ij)} = d_{kj}$$

if $d_{ki} < d_{kj}$, then

$$|d_{ki} - d_{kj}| = d_{kj} - d_{ki}$$

and therefore

$$d_{k(ij)} = d_{ki}$$

i.e. the formula gives $d_{k(ij)} = \min\{d_{ki}, d_{kj}\}$, as required.

B.5   Chapter 7

7.1

$$\log m_{ij} = \log N + \log p_{i.} + \log p_{.j} \tag{1}$$

since $m_{i.} = Np_{i.}$ and $m_{.j} = Np_{.j}$, this may be written as

$$\log m_{ij} = \log m_{i.} + \log m_{.j} - \log N \tag{2}$$

Summing (2) over i we have

$$\sum_{i=1}^{r} \log m_{ij} = \sum_{i=1}^{r} \log m_{i.} + r \log m_{.j} - r \log N$$

Summing (2) over j we have

$$\sum_{j=1}^{r} \log m_{ij} = c \log m_{i.} + \sum_{j=1}^{c} \log m_{.j} - c \log N$$

Finally summing (2) over i and j we have

$$\sum_{i=1}^{r} \sum_{j=1}^{c} \log m_{ij} = c \sum_{i=1}^{r} \log m_{i.} + r \sum_{j=1}^{c} \log m_{.j} - rc - \log N.$$

By setting

$$u = \frac{1}{rc} \sum \sum \log m_{ij},$$

$$u_{1(i)} = \frac{1}{c} \sum_{i=1}^{c} \log m_{ij} - u,$$

and

$$u_{2(j)} = \frac{1}{r} \sum_{i=1}^{r} \log m_{ij} - u,$$

it is now a matter of simple algebra to show that equation (2) may be written in the form

$$\log m_{ij} = u + u_{1(i)} + u_{2(j)}$$

7.3

$$\log m_{ijk} = u + u_{1(i)} + u_{2(j)} + u_{3(k)} + u_{12(ij)}$$
$$+ u_{13(ik)} + u_{23(jk)} + u_{123(ijk)} \quad \text{(saturated model)}$$
$$\log m_{ijk} = u + u_{1(i)} + u_{2(j)} + u_{3(k)} \quad \text{(minimal model)}$$

B.6   Chapter 14

14.1 The predicted covariance matrix is found from

$$\Sigma = \Lambda_x \Phi \Lambda_x'$$

leading to

$$\Sigma = \begin{pmatrix} \lambda_{11}^2 + \theta_{11} \\ \lambda_{21}\lambda_{11} & \lambda_{21}^2 + \theta_{22} \\ 0 & 0 & \lambda_{32}^2 + \theta_{33} \\ 0 & 0 & \lambda_{42}\lambda_{32} & \lambda_{42}^2 + \theta_{44} \\ 0 & 0 & \lambda_{53}\lambda_{32} & \lambda_{52}\lambda_{42} & \lambda_{52}^2 + \theta_{55} \\ 0 & 0 & \lambda_{62}\lambda_{32} & \lambda_{62}\lambda_{42} & \lambda_{62}\lambda_{52} & \lambda_{62}^2 + \theta_{66} \\ 0 & 0 & \lambda_{73}\phi_{32}\lambda_{32} & \lambda_{73}\phi_{32}\lambda_{42} & \lambda_{73}\phi_{32}\lambda_{52} & \lambda_{73}\phi_{32}\lambda_{62} & \lambda_{73}^2 + \theta_{77} \\ 0 & 0 & \lambda_{83}\phi_{32}\lambda_{32} & \lambda_{83}\phi_{32}\lambda_{42} & \lambda_{83}\phi_{32}\lambda_{52} & \lambda_{83}\phi_{32}\lambda_{62} & \lambda_{83}\phi_{32}\lambda_{72} & \lambda_{83}^2 + \theta_{88} \end{pmatrix}$$

The model is **not** identified.

Index